EXPORTING PARADISE

TOURISM AND DEVELOPMENT IN MEXICO

TOURISM SOCIAL SCIENCE SERIES

Series Editor: Jafar Jafari
Department of Hospitality and Tourism, University of Wisconsin-Stout, Menomonie WI 54751, USA.
Tel: (715) 232-2339; Fax: (715) 232-3200; E-mail: jafari@uwstout.edu
Associate Editor (this volume): Oriol Pi-Sunyer
University of Massachusetts-Amherst, USA

The books in this Tourism Social Science Series (TSSSeries) are intended to systematically and cumulatively contribute to the formation, embodiment, and advancement of knowledge in the field of tourism.

The TSSSeries' multidisciplinary framework and treatment of tourism includes application of theoretical, methodological, and substantive contributions from such fields as anthropology, business administration, ecology, economics, geography, history, hospitality, leisure, planning, political science, psychology, recreation, religion, sociology, transportation, etc., but it significantly favors state-of-the-art presentations, works featuring new directions, and especially the cross-fertilization of perspectives beyond each of these singular fields. While the development and production of this book series is fashioned after the successful model of *Annals of Tourism Research*, the TSSSeries further aspires to assure each theme a comprehensiveness possible only in book-length academic treatment. Each volume in the series is intended to deal with a particular aspect of this increasingly important subject, thus to play a definitive role in the enlarging and strengthening of the foundation of knowledge in the field of tourism, and consequently to expand its frontiers into the new research and scholarship horizons ahead.

Published and forthcoming TSSSeries titles include:

JÓZSEF BÖRÖCZ (Rutgers University, USA)
Leisure Migration: A Sociological Study on Tourism

DENNISON NASH, (University of Connecticut, USA)
Anthropology of Tourism

PHILIP L. PEARCE, GIANNA MOSCARDO & GLENN F. ROSS (James Cook University of North Queensland, Australia)
Tourism Community Relationships

TREVOR SOFIELD (Murdoch University, Australia)
Empowerment and Sustainable Tourism

BORIS VUKONIĆ (University of Zagreb, Croatia)
Tourism and Religion

NING WANG (Zhongshan University, China)
Tourism and Modernity: A Sociological Analysis

Related Elsevier journals
Annals of Tourism Research
Cornell Hotel and Restaurant Administration Quarterly
International Journal of Hospitality Management
International Journal of Intercultural Relations
Tourism Management
World Development

EXPORTING PARADISE

TOURISM AND DEVELOPMENT IN MEXICO

MICHAEL CLANCY

University of Hartford, USA

2001

Pergamon
An Imprint of Elsevier Science

Amsterdam – London – New York – Oxford – Paris – Shannon – Tokyo

ELSEVIER SCIENCE Ltd
The Boulevard, Langford Lane
Kidlington, Oxford OX5 1GB, UK

First edition 2001

Library of Congress Cataloging in Publication Data
A catalog record from the Library of Congress has been applied for.

British Library Cataloguing in Publication Data
A catalogue record from the British Library of Congress has been applied for.

ISBN 0-08-043715-X

⊚ The paper used in this publication meets the requirements of ANSI/NISO Z39.48-1992 (Permanence of Paper).
Printed in The Netherlands.

Contents

Acknowledgements

Many debts have been incurred while writing this book. Special thanks go to Michael Barnett at the University of Wisconsin, who aided in this project under unusual circumstances. Michael always pushed for more and as a result, the book is better for it. Thanks also to Leigh Payne, Barbara Stallings, Bill Thiesenhusen, and the late Charlie Gillespie. The research was funded in part by grants from the Tinker foundation and the graduate school at the University of Wisconsin.

In Mexico, the author benefited from the aid of numerous people. Special thanks go to Adalberto Garcia Rocha, then chair at the Center for Economic Studies at El Colegio de Mexico, who arranged for research privileges at the fine library at the university. Thanks also to Daniel Hiernaux, who provided great guidance on tourism in Mexico. Others as well were guides to key documents and interviews. Officials at FONATUR and especially SECTUR were welcoming and provided access to many official documents in libraries and archives. They also spoke quite candidly about the politics of tourism development in Mexico. The same may be said of representatives in the private sector in Mexico and the United States.

Various drafts of the book received many readings along the way. Special thanks go to Mary Geske and Greg White, who read countless versions. Oriol Pi-Sunyer served as an associate editor of the final manuscript and his careful readings and suggestions of several drafts make it a much better book. Others who have either read portions of the manuscript or heard some of the central arguments included in it include Patrick Barrett, Sam Crane and Jim Mahon. Thanks also to the series editor, Jafar Jafari, and David Lamkin at Elsevier.

Finally, special thanks go to Mary and the author's son, Patrick. Both have supported this project through its long duration in countless ways. The book is dedicated to them.

Chapter 1

Introduction: Tourism, Industrial Change and Development

Forty years ago few international tourists knew Mexico. Among the few thousand who did, some journeyed to the distant beach resort of Acapulco but most confined themselves to the border regions of the country, the northwest Pacific coast below Tijuana, and to Mexico City. Today some 20 million international tourists visit Mexico each year, injecting more than $8 billion into the local economy. Among the remarkable aspects of this transformation is that it did not happen by accident. Instead, a group of developmentally minded technocrats within the Mexican government planned much of the tourism growth in the name of development. In this sense their plan was to "export paradise". In other words, they sought to take advantage of Mexico's beaches, cities, ruins, people, and weather — the very things that tourists seeking leisure activities enjoy — and utilize them as a development tool. By "exporting" tourism services to international tourists, they could create jobs, promote regional development, and earn foreign exchange.

Export-led growth strategies have recently become attractive to many practitioners and analysts of Third World development. Undoubtedly much of the enthusiasm associated with export promotion derives from the East Asian experience of postwar Japan and later the so-called "four tigers" of South Korea, Taiwan, Hong Kong and Singapore, at least before the onset of the 1997 Asian currency crisis. In addition, the lost decade of development that much of the rest of the Third World experienced in the 1980s led to disaffection with alternative strategies. As countries turn to greater integration into the world economy as a means toward development, however, what might one expect? What are the links between export promotion and developmental outcomes?

To be sure, this is a big and complex question. Defining export promotion is difficult and adding development invites a stroll through a particularly treacherous minefield. Despite these dangers, however, attempting to understand the connections between the two is both possible and necessary. Careful specification of both concepts may lead to a workable study that at least begins to get at an answer. Another caveat is that the answer is not likely to be a simple one; in fact the most obvious response is that "it depends". The question is upon what. Not all export promotion is the same and neither are all exports.

This leads into the two primary tasks of this study. One is to provide an analytical framework for understanding the specific factors that determine the developmental consequences of export orientation. A central facet of the approach here is to consider both cause and consequence of development patterns at the industry or sectoral level. In other words, examining "development" is best done at the level of industry. What is more, while integrating state and systemic variables into the framework, the argument is that much of the explanatory power over developmental outcomes lies with aspects of the industry itself. Once the skeleton of the basic framework is laid out in Chapter 2, it is then utilized in order to carry out the second task, examination of an early case of export promotion, Mexican tourism.

The case deserves attention for several reasons, and while a more detailed argument is provided below, a few words of justification are in order here. Tourism deserves greater empirical attention due to its sheer size in the world economy and within the developing world. Roughly one in every four international tourist dollars is spent in the Third World (WTO, 1997). Tourism invites a great deal of controversy in development studies, but whether aiding or retarding development in Mexico its importance to the economy demands greater attention. The activity makes up a significant portion of the overall economy, is a leading source of foreign exchange, and provides a large number of jobs.

In addition, tourism in Mexico was promoted comparatively early, dating back to the late 1960s. The export of tourism — defined as the provision of tourist-related services to foreigners visiting the country — constitutes one of the few cases of non-primary product export promotion in Latin America dating back that far. The obvious advantage is that the long-term outcome of export promotion may be analyzed.

A third rationale is the need for greater empirical attention to services within development studies. Services, which account for more than half of world GDP and roughly one fifth of world trade (Riddle, 1986; Orr, 1992), lie at the fringes of development studies. Tourism, the largest service industry in the world and the second largest single item of world trade, is a logical choice in this sense, and Mexico is especially appropriate because it possesses one of the largest tourism export sectors in the Third World (WTO, 1998).

Further, studying Mexican tourism as a case of export promotion and development is also consistent with the idea that development is best understood at the industry level. Development writ large has become a fuzzy concept. Little agreement exists on its definition or how to operationalize it. One alternative is to disaggregate and simplify development by studying the political economy of change within key industries and economic activities. Moreover, the factors that affect developmental outcomes also tend to cluster around individual industries. Most state policies, for instance, are made at the industry level. Lastly, a central contention made here is that developmental outcomes also vary *due to* the nature of the industry itself. The obvious danger here is to claim that "A causes A". As the discussion below and in the next chapter makes clear, however, one may carefully distinguish the causes and consequences. In short, a more micro view is warranted.

The remainder of this chapter provides greater detail on the nature of this study. It includes an elaboration of terms and locates the study of Mexican tourism within the larger contemporary debate on development. The next chapter lays out in detail a framework for understanding the variables and contingencies for analyzing export promotion and development outcomes.

Development as Industrial Transformation

Development is by its nature value laden and wedded to modernism. As Escobar (1995), Sachs (1992), and others have pointed out, its invention as a field of study after World War II presumed an existing oppositional state to be escaped from. Much of this imagination has been binary in nature, reducing traditional societal practices and structures as "underdevelopment" and prescribing remedies that would allow people to somehow become "modern" (Shrestha, 1995). This critique is not to deny that poverty and hunger truly exist, but rather to simply suggest that the transformative project associated with development studies and practice involves a particular "imagining" of the problem that is tied to ideology at least as much as science. Additionally, to the extent that many if not most development problems and solutions are conceptualized by political and economic elites, that conceptualization itself as well as the actions that stem from it become an exercise in raw power. Certainly much of the story of import substitution and export promotion involves international and domestic development elites who impose *their* solutions on *their* conceptions of the development "problem".

This study examines exactly such a case. As such, much of the task here is simply to tell the story of which problems were identified and how they were conceptualized, by whom, what solutions were put in place, and what outcomes resulted. Any work on development, however, must specify *which* outcomes are important. Recent works in political economy have attempted to closely examine industrial change within Third World countries. Gary Gereffi (1990a, 1989), for example, asks three questions regarding the evolution of what he calls development patterns. One, what are the most prominent economic activities historically within a country? Two, to what extent are they inwardly or outwardly oriented? Three, who are the primary economic agents relied upon in these industries? Gereffi's work is similar to others who eschew even the term "development" in favor of "development patterns" (1990a), "structural change" (Evans, 1992; 1995) or "industrial change" (Haggard, 1989). The approach here is somewhat akin, although more micro-oriented. This study consciously selects export activities, and subsequently provides a more thorough examination of one such leading industry. In this manner it is consistent with Gereffi's (1983) earlier work as well as recent work on global commodity chains, which center upon individual industries and their global organization (Gereffi & Korzeniewicz, 1994; Clancy 1998).

Structural changes that take place within the economic activity over time may best be thought of as industrial transformation. Industrial transformation may be defined simply as growth plus attention to the distribution of material benefits deriving from that growth. In this sense the outcome to be explained can be summarized in two questions: what kinds of growth levels are achieved under export promotion (and why), and who are the primary beneficiaries from such growth (and why)? This significant change at the industry level as a measure of development is rather modest, yet may easily be made operational.

Growth is easily examined at the industry level under export promotion. Do exports increase over time, and by how much? For tourism the dual measures of foreign arrivals and export earnings serve this purpose. Additional measures such as job creation and contribution of the industry to the overall economy augment these primary indicators and demonstrate the importance of the activity to the economy as a whole. These indicators are tied to most traditional conceptions of development. The primary beneficiaries within the activity, the distributional question, are considered through answering a modified version of Gereffi's third question, that of the economic agents controlling the industry. For Gereffi the key distinction to be made is ownership of leading firms by the state, domestic business, or transnational corporations (TNCs). Because ownership may only be part of the story, however, broader strategic and contractual relationships among these actors are highlighted here. The task then is to uncover changing patterns of ownership and control within the economic activities that make up tourism. Emphasis is placed on two of the most important industry sectors, hotels and airlines. This is common in tourism studies, although some also add a third sector, that of tour operators (Britton, 1981, 1982; Lea, 1988). Attention to patterns of ownership and control offers a rough measure of who the beneficiaries are and provides a more direct and accurate account of control.

To be sure, industrial transformation as defined here constitutes a somewhat narrow conceptualization of development. Aside from the obvious simplicity, however, there are other advantages to such an approach. First, the components are intimately tied to broader aspects of development such as wealth creation and equity. Second, the outcomes of interest here are derived from contemporary development theories. Most strands of development theory are concerned with both growth and equity, but emphasis in evaluating developmental outcomes tends to vary. Briefly, for years dependency, world-system, and other neo-Marxist theorists have contended that the manner in which Third World countries have been inserted into the world economy has placed significant limits on both growth prospects and especially on broad-based, relatively equitable economic development. In short, peripheral development tends to go in stop-go cycles and results in development for the few (Biersteker, 1987; Cardoso & Faletto, 1979; Evans, 1979; Frank, 1967, 1969). While growth is important, development *for whom* is emphasized. Gereffi's third question of identifying the primary economic agents relied upon in leading industries, for instance, almost certainly stems from earlier dependency theory concerns over the denationalization of key industries by foreign capital (Cardoso & Faletto, 1979; Evans, 1979).

The orthodox response is that factors other than peripheral status have led to such negative outcomes. The most commonly found culprits have been import substitution industrialization, related protectionism and heavy state intervention in the economy (Balassa, Bueno, Kuczynski & Simonsen, 1986; Haggard, 1990). What then should export promotion promise? The answer is not completely clear. Many economic liberals would expect that export "orientation" in the form of trade openness leads to competitiveness and the efficient allocation of resources. Therefore, it offers the best chance for development, measured mainly by aggregate growth levels, employment creation, and the generation of export earnings. Certainly growth appears to be first among equals, however, as many other benefits are derivative of an expanding pie. The question of equity is more muddled. Cases such as South Korea and Taiwan suggest that growth with a relatively broad distribution of benefits is possible (World Bank, 1994). On the other hand, recent and continued economic stagnation within much of the Third World has led to a renewed emphasis on growth as a first priority (Balassa *et al.*, 1986).

Export Promotion

While growth is a concern for most who are interested in development, its connection to export promotion deserves further discussion. Initially export-led growth must be defined. There is no one form of export promotion or orientation posited here; instead it is best thought of as existing within a continuum. Export strategies originate as state policies that may range from outward orientation at one end of the continuum to an export "push" on the other. The minimalist outward orientation amounts to a set of policies that do not create biases against exports. Primarily market oriented, they are most obvious in exchange rate values that are neutral or slightly undervalued. Policies favoring export orientation are most commonly made at the economy-wide level but also may be directed at individual industries.

The other end of the range, the export push, constitutes a much more activist policy of encouraging outward trade of goods or services. In addition it usually refers to a targeted area of the economy. Specifically, state managers frequently identify and "push" exports through the selective granting of benefits or the threat of punishments to producers and traders. The key feature is that price signals are thus not market conforming. In the case of South Korea, for example, state managers created preferential financing and tax schemes for those who met export quotas in targeted areas of the economy. Those who did not forfeited future access to such benefits (Bradford, 1990; Haggard, 1990; Haggard & Moon, 1989). Further, state-owned firms may take on some of the roles associated with the export push, ranging from finance to research and development to direct production.

The arguments in favor of some form of export promotion are rather straightforward, and three in particular stand out. The first is more theoretical or logical in nature: export promotion should benefit development simply through introducing greater competitive pressures to industry or the economy as a whole. International market competition quickly weeds out inefficiencies, thereby leading to higher productivity, better allocation of resources, the spread of technology, and even entrepreneurial learning. This is even more likely when export promotion is accompanied by liberalization of home markets. Together these should also reverse or preclude the rent-seeking activities often associated with import substitution. Export orientation may also overcome insufficient domestic demand, and foreign exchange earnings may even be diverted to strategic uses elsewhere in the economy.

The next arguments rely on empirical cases and studies. Second is the aforementioned historical example of East Asia. While export promotion schemes have varied, clearly the developmental record within the region over the past 30 years is the envy of much of the Third World. Average annual GDP growth rates for East Asia were 8.7 percent between 1970 and 1990. In comparison Latin America grew by an average of 3.75 percent and sub-Saharan Africa by 2.7 percent (Stallings 1995:24). Third, empirical studies using quantitative methods have found a positive correlation between export growth and overall growth in gross domestic product (Bhagwati, 1978; Ram, 1987). Some question has arisen with regard to causation as well as identification and specification of export orientation (Barham, Clark, Katz & Schurman, 1992), but again the general lesson taken from such arguments is that export growth contributes to certain desirable development outcomes.

In a classic statement following the onset of the debt crisis in the early 1980s, Balassa and colleagues used portions of all of these arguments to lobby for a sustained and radical shift toward what they call "outward orientation" in Latin America. The cornerstone of this new model was to be market-based export promotion. They contended:

> Only such a focus will enable the countries of the region to achieve self-sustaining growth and simultaneously to service their external debt, because only such a strategy will generate both the needed foreign exchange and the essential stimulus to domestic production (Balassa *et al.*, 1986:24).

More recently the World Bank has been at the forefront of encouraging export orientation. The Bank has based most of its arguments on the relationship between exports and growth, but recently has also prescribed exports from a distributional standpoint, arguing that export-led economies enjoy greater equity and that such strategies particularly benefit workers. In fact, the Bank's 1995 version of *World Development Report,* is devoted to workers. It contends that "[m]arket-based development, which encourages firms and workers to invest in physical capital, new technologies, and skills, is the best way to deliver growth and rising living standards for workers" (World Bank, 1995:3).

Indeed, a large number of countries in not only Latin America but the Third World in general appear to be heeding such advice. While in some cases it may be argued that policies are imposed from outside, the sheer size and scope of the change of the shift suggests considerable enthusiasm for outward orientation within several countries. Many such strategies are market oriented, suggesting more of an export orientation rather than export "push" (Biersteker, 1995; Bradford, 1990).

Export Promotion and Industry Studies

Why the need for further study of export promotion, especially at the industry level? The most direct response is that previous studies are controversial and have run into significant barriers. Export growth may be an effect rather than the cause of economic growth, or both may be affected by a third factor. The question with regard to East Asia is whether a generalizable model may be derived or if, in fact, it is best understood as a developmental anomaly. Cumings (1987), for instance, argues the dynamism is largely the result of a unique set of regional historical circumstances. On a related point, there are few long-term cases of export orientation to study outside East Asia. This is especially true in Latin America, where most experiments in liberal or outward-oriented development prior to the 1980s were short lived (Balassa, *et al.*, 1986).

Industrial transformation as defined above includes growth and change, but also inquires about the distribution of benefits. These are best analyzed at the industry level, especially when export promotion takes place at that level. Certainly multiplier and other spillover effects will be felt from dynamism in any industry, but the direct beneficiaries of export promotion are best found at the industry level and through analyzing ownership and control. Additionally, as Chapter 2 argues in detail, because not all industries are the same, the outcomes of industrial transformation will likely vary.

Tourism, and particularly the Mexican version, constitutes an ideal case for analyzing the developmental outcomes of export promotion. Several advantages are apparent. One mentioned above is the need for attention to a badly neglected area in development studies: services. Tourism constitutes a service export even though it is foreign tourists who cross borders to "import" tourism services. Services have in the past frequently been treated as a residual category even though they tend to make up roughly one-third to one-half of Third World economies. In part this neglect of services is due to what Bhagwati (1987) calls a "haircut" view of services. The mistaken assumption is that since one cannot get a haircut long distance, then services in general are not tradeable. Although this is clearly not the case with many services, debates on defining and measuring services and trade continue (Gibbs, 1987; Riddle, 1986:113–116; Snape, 1990). Indeed, services have enjoyed a boom in international trade in recent years. Between 1978 and 1990 the value of commercial services in world trade grew from $50 billion to $770 billion (Madeley, 1992:72) and while trade of goods grew by 74 percent from 1981 to 1991, trade in services jumped by 230 percent (Kenen, 1994:262). Most trade in services takes place between wealthier countries, but its importance in the Third World is growing.

Tourism is said to constitute the largest service industry in the world and it also accounts for the single largest item in international trade of services (C. Richter, 1987; L. Richter, 1989:3). Recent data (Table 1.1) show *international* tourism was a $445 billion business by 1998. As the table also shows, global tourism has grown rapidly in recent decades, not only among wealthier countries, but also in the Third World. Linda Richter (1989:3) reports that by the end of the 1980s more than 125 nations considered tourism to be a major industry and several have integrated this exports into a central part of a larger development strategy. The most obvious cases are small island nations, but in addition larger nations such as Mexico, China, Kenya, Egypt, and Thailand have actively promoted tourism exports (Hoekman, 1990).

Table 1.1: World Tourism Arrivals and Receipts.[a]

Year	Arrivals (in thousands)	Variation %	Receipts ($ Million)	Variation %
1950	25,282	–	2,100	–
1960	69,320	10.6	6,867	12.6
1965	112,863	10.3	11,604	11.1
1970	165,787	8.1	17,900	9.1
1975	222,290	6.1	40,702	18.0
1980	285,997	5.2	105,320	21.1
1985	327,188	2.8	118,084	2.4
1986	338,854	3.6	143,475	21.5
1987	363,766	7.4	176,795	23.2
1988	394,810	8.5	204,290	15.6
1989	426,461	8.0	221,263	8.3
1990	458,229	7.5	268,928	21.5
1991	463,951	1.3	277,568	3.2
1992	503,356	5.5	297,853	13.2
1993	500,142	8.5	315,103	13.5
1994	550,471	6.1	353,998	9.2
1995	565,495	2.7	405,110	14.4
1996	596,524	5.5	435,594	7.5
1997	610,763	2.4	435,981	0.1
1998	625,236	2.4	444,741	2.0

[a]WTO (1999).

The Mexican case is particularly appropriate because this country has one of the largest tourism export sectors in the Third World. By 1996 the country drew more than 40 percent of all international tourists to the Western Hemisphere outside the United States and Canada (WTO, 1998) and ranked seventh in the world in popularity (WTO, 1998). The numbers are perhaps even more impressive when considered over time: Mexico drew 2.2 million tourists in 1970, raising that figure by nearly three times in 1991. During the same period foreign exchange earnings from tourism increased more than ninefold, from $415 million to $3.8 billion (SECTUR nd). As Table 1.2 summarizes, international tourists and receipts increased steadily over time. From 1970 to 1991 arrivals grew by an average annual rate of 5.1 percent while receipts grew by 11.1 percent (SECTUR, 1992). Because the government changed its method of counting after that point to conform with World Tourism Organization methodology, subsequent figures are not comparable, but by 1998 Mexico attracted 19.8 million international tourists (*Travel and Tourism Intelligence*, 1999).

This growth was not without occasional downturns — most notably after the oil shocks of 1974–1975 and 1981–1982 and a travel boycott sponsored by pro-Israeli groups after Mexico supported a United Nations vote that equated Zionism with racism — but the record over the long term is impressive. Export receipts from tourism outpaced both GDP growth and exports as a whole during the 25-year period, during which tourism has become increasingly important within the overall Mexican economy. It makes up the largest service activity in the country both in terms of share of GDP and employment. Throughout much of the 1980s and early 1990s tourism served as the second largest source of foreign exchange behind oil. Some estimates suggest that in the late 1980s more than seven percent of the workforce was employed directly or indirectly in the industry (Enríquez Savignac, 1988), while others range to as high as nine percent (SECTUR, 1992).

This outcome is the result of conscious export promotion. Tourism did not simply come to Mexico due to geography or topography. These were important, but not decisive. Instead, beginning in the late 1960s state policy makers openly targeted the industry for expansion and it underwent an "export push". The state subsequently oversaw the creation of five new resorts during the 1970s and 1980s. Well known resorts such as Cancún and Ixtapa were transformed from sparsely populated areas where residents were devoted largely to subsistence agriculture into world-class destinations. Equally important, government officials appear committed to placing tourism at the leading edge of an export-led growth strategy today and in the future.

Table 1.2: Mexican Tourism Arrivals and Receipts, (1970–1998).[a,b]

Year	Arrivals (in millions)	Variation (%)	Receipts (in billions)	Variation (%)
1970	2.25	–	.415	–
1971	2.51	11.5	.461	11.1
1972	2.92	16.2	.563	22.0
1973	3.23	10.7	.724	28.7
1974	3.36	4.2	.842	16.3
1975	3.22	–4.3	.800	–5.0
1976	3.11	–3.4	.836	4.4
1977	3.25	4.5	.867	3.7
1978	3.75	15.6	1.121	29.4
1979	4.13	10.1	1.443	28.8
1980	4.14	0.2	1.671	15.8
1981	4.04	–2.6	1.759	5.3
1982	3.77	–6.7	1.405	–20.1
1983	4.75	26.1	1.625	15.5
1984	4.66	–2.0	1.953	20.2
1985	4.21	–9.6	1.720	–11.9
1986	4.63	9.9	1.792	4.2
1987	5.41	16.9	2.274	26.9
1988	5.69 (14.14)	5.3	2.544 (4.0)	11.9
1989	6.19 (14.96)	8.7	2.954 (4.7)	16.1
1990	6.39 (17.18)	3.3	3.401 (5.5)	15.1
1991	6.37 (16.28)	–0.3	3.783 (5.9)	11.3
1992	– (17.27)	(6.8)	– (6.0)	(1.6)
1993	– (16.44)	(–3.1)	– (6.2)	(1.4)
1994	– (17.18)	(4.5)	– (6.4)	(3.2)
1995	– (20.24)	(17.8)	– (6.2)	(–2.9)
1996	– (21.40)	(5.7)	– (6.9)	(12.2)
1997	– (19.35)	(–9.6)	– (7.6)	(9.5)
1998p	– (19.81)	(2.4)	– (7.9)	(4.0)

[a]Figures in parentheses reflect a revised methodology for counting arrivals and receipts in accordance with the WTO. Previously Mexico only counted international tourists arriving by air. The revised figures reflect arrivals by land and air but only count those travellers who stay in the country for more than 24 hours.
[b]SECTUR (1992); WTO (1999); Travel and Tourism Intelligence (1999).
[p] = preliminary data.

Conclusion: Evaluating Tourism as Export Promotion

On what basis should the outcome of tourism as export promotion in Mexico be evaluated? This is a difficult question, and one that will be reintroduced in the concluding chapter. Tourism is uniquely controversial in studies of Third World development. Analysts have long been divided over the potential benefits international tourism may provide for development (Britton, 1981, 1982; Harrison, 1992; Lea, 1988; Pearce, 1989; Turner & Ash, 1975). On the one hand, the industry creates employment, export earnings, and tax revenue, and establishes linkages with other economic activities. On the other, these benefits may frequently be overstated. Employment created by the industry is often seasonal and low paying. Export earnings fluctuate by the degree to which local linkages are established. This is often difficult as many imports stem from the tastes and consumption patterns of foreign tourists. Foreign investment in the industry is usually quite heavy and high repatriation of profits and foreign employment have been reported, leading many to question who benefits from tourism activities.

Tourism is not a developmental savior, but it is not necessarily inherently exploitative. One may proceed more inductively, but still required are guidelines for evaluation. In the first instance one could judge outcomes on the terms set forth by those who created the export promotion scheme. What were policymakers trying to accomplish, and were they successful? In Mexico, state policymakers held two overriding concerns; increasing export earnings and enhancing regional development by creating economic opportunities in some of the poorest regions of the country. Based on these criteria the tourism export push was successful on its own terms, as is shown in the next chapters.

Evaluating development outcomes may also go beyond the intent of policymakers, however, by suggesting other basic criteria, which is done above with industrial transformation. This first requires documenting growth and patterns of ownership and control as the *outcome* of export promotion, and then utilizes an explanatory model in order to account for this outcome. Such a model is presented in the next chapter, and it basically constitutes a framework for explaining the linkages between export promotion and the developmental outcomes noted above. Although it is then utilized to examine the case of Mexican tourism, it also may be applied to study tourism elsewhere as well as other cases of export promotion in the Third World.

Chapter 2

Theoretical Issues: Explaining Industrial Transformation

For those interested most in the *growth* aspect of export-led strategies, Mexican tourism over the past quarter century represents a success story. According to standard methodology, export earnings in 1992 were 17 times what they had been in 1970. Tourism growth easily outpaced expansion of the overall Mexican economy during this period and was one of the single most dynamic activities within the country. Growth, as noted in the previous chapter, is only one of two components that make up industrial transformation, however. The second is distribution of material benefits accruing within dynamic industries, which is measured by examining patterns of ownership and control. Here, as subsequent chapters demonstrate, the outcome of tourism promotion is more mixed. It is important to note that the industry has not been "captured" by transnational corporations (TNCs), as many critics of Third World tourism would contend. Foreign ownership and control, however, is significant and where foreign TNCs do not dominate, home grown ones do. In short, tourism has become big business in Mexico and big business has come to largely control the industry.

In order to understand how this transformation came about, a broader analytical framework is needed. The task is to specify the manner in which various factors associated with export promotion shape specific development patterns. Frequently mainstream development studies treat the relationship between government policy and development outcomes in a simplistic manner: bad policy leads to bad development outcomes and good policy to good. A more accurate approach recognizes that export orientation/promotion does not inherently lead to positive or negative developmental outcomes. Instead various structural attributes along with choices made shape resulting development patterns. In short, not only can the forces behind industrial transformation for tourism in Mexico be identified, but so can the larger set of generalizable determinants associated with cases of export promotion.

The framework presented in this chapter amounts to a two-step model that combines structural features and strategic choices. Within the initial step two factors are important. First is the stimulus that facilitates the decision to promote or orient toward exports. Where, in other words, are the seeds of export promotion/orientation sown? Second, and stemming from above, is the role of state actors in responding to this stimulus. How and in what form is the specific export scheme undertaken influenced by factors internal to the state? The second step of the model contains industry characteristics that shape growth but especially the distributional features of industrial transformation. The nature of these elements and their relationship to one another are further developed here.

Promoting Exports

Industrial transformation is the product of both impersonal structures associated with global and local markets and governing structures and calculated choices made by individuals within government agencies and firms. In addition, specific attributes of the industry and even sub-industries in question raise further contingencies. In short, industrial transformation is neither purely structurally determined nor the matter of simple choice. Because the interest here is on export promotion and industrial transformation, attention must first be given to the context and choice of export promotion. This first step of the model includes the environment stimulus leading to the decision to export.

The developmental outcomes resulting from export promotion/orientation are in part a function of the very nature of that promotion. Because of this one first needs to know what conditions surround the decision to export. The most obvious factor is the set of general macroeconomic conditions facing policymakers because the decision to promote exports frequently takes place as a response to changing conditions within that environment. This is not to suggest that other stimuli could not also be at work: drought, famine, wars, population or migration pressures, a new government, and other factors could all lead to policy change that has this effect. The focus here is on the economic environment because general economic conditions motivate state officials to act in specific ways.

For present purposes two particular contingencies are important. First, broad features associated with international markets have a varying impact on the domestic economy over time, thereby changing the degree to which state actors have room to maneuver. An obvious example can be seen with the events leading up to and following the Third World debt crisis. Recycled petrodollars created the availability of cheap, easily-obtained loans from private banks in the

1970s and broadened the range of policy options for state actors. The reverse was true, of course, after the onset of the debt crisis in 1982. Therefore, market factors directly affect the amount and type of resources available to policy-makers and also send signals that affect their decision-making calculus. In short, as Stallings suggests, if national development is "embedded" in a larger international context, so are the economic policy decisions made by state managers (1978:12).

Second, the interaction between markets and state response has something of a cumulative component to it. In other words, historical patterns of state responses to changing international market conditions create structures and incentives that shape future challenges and choices. Adoption of import-substi-tution policies in Latin America, for example, most commonly came about as an *ad hoc* response to the twin shocks of the Great Depression and World War II. As industrialization proceeded and an import substitution coalition was formed and became more entrenched, however, policy options favoring alternative development strategies became more difficult to implement. In fact this forms one explanation for the failure of Latin American countries to adopt broad-based export-oriented strategies at the same time East Asian countries did (Kaufman, 1990). Haggard (1990) contends that societal group pressures were one of several overlapping factors in the cases of Brazil and Mexico. The larger point is that the broad political economy affects state actors' motivation and ability to act. It provides a contextual setting that aids in understanding incen-tives for state action.

In addition to motivation for state action, macroeconomic factors tend to determine the resources available to the state (Haggard, 1990:29–33). Resources frequently shape not only whether state action is undertaken but also the form of that action (Evans, 1995:77–81). State policies may range from distant interest in industrial development to much more costly provision of public goods. The state may also offer subsidies or engage in production itself. Here, then, the availability of material resources likely affects *which* policy instruments are chosen. At the very least it will preclude certain ones.

This is not to argue that the environment alone is sufficient for explaining policy change. Historically, nearly identical macroeconomic stimuli such as international oil shocks or the debt crisis led to varying responses among Third World policymakers. On the other hand, attention to the *nature* of such stimuli is important. One expectation is that the intensity of the stimulus to export will likely affect the form of export promotion/orientation chosen by policymakers. Again a caveat is in order: this is not to suggest a clear and direct relationship between the nature of the shock and policy choice, but instead a general tendency where macroeconomic "shocks" are likely to lead to more market orientation in the promotion of exports, while less intense stimuli occurring over a longer time period, because they offer policymakers more room to maneuver, frequently lead to selective, nonmarket-conforming export pushes.

The reasons for this are fairly straightforward. Shocks frequently demand quick and decisive action from state managers. Market-conforming initiatives that have the effect of promoting exports — most commonly currency devaluation and the elimination of export taxes — tend to be made at the economy-wide level where the effect will be most significant and immediate. In addition, because macroeconomic shocks most commonly take the form of balance of payments crises, the measures are doubly effective because they affect both export and import prices. Further, while the decision to promote exports through market orientation may involve significant political costs, there are reasons to believe that it is most likely to take place under conditions of crisis or heightened pressure. State managers may be able to deflect criticism by societal groups to foreign multilateral actors such as the IMF, World Bank, foreign aid donors, or advisors (Williamson & Haggard, 1994). In addition, while a "blocking" political coalition in support of previous policies may exist, it may in reality be weaker than it initially appears to be, especially if benefits to societal groups from the new set of policies are felt early on. Waterbury (1992:192–193), for instance, contends that opposition to privatization is frequently feeble because it is quickly discredited and is severely weakened by the withdrawal of political support by the state. The new winners created by privatization will provide an alternative basis of political support for the regime. One would expect the same argument to hold in the case of trade liberalization.

Similar reasoning also holds in the case of stimuli favoring an export push. A push will most likely originate in an environment that provides some stimulus for export growth, but does not constitute a crisis. Export pushes frequently target individual, often new economic activities and are carried out by manipulating prices and other incentives. As a result, benefits deriving from the industries in question are likely to be too small — at least initially — and arrive too late in order to aid in responding to economic crises. In fact in the short term, because such programs involve altering prices, the state usually offers material incentives such as subsidies or reduced tax revenue. Under conditions of macroeconomic crisis the state seldom possesses the necessary resources with which to implement such a program.

Of course most stimuli for export-led growth strategies are not easily identifiable as constituting a crisis or non-crisis; many will fall somewhere in between. For the analyst the task is to uncover historical conditions that prevailed at the time export projects were adopted. Chapter 3 thus presents a brief profile of the evolution of the post-war Mexican political economy leading up to the late 1960s, when the decision to promote tourism was adopted. The expectation is that its promotion, because it took the form of a nonmarket-oriented push, originated in a macroeconomic environment that demonstrated a long-term need for export revenues but did not constitute a crisis. The treatment emphasizes Mexico's external sector leading up to the initiation of the tourism push, focusing on the ISI model of industrialization, the changing composition of Mexican trade and more general balance of payments trends over time. Other internal factors, such as demographic pressures, evolving socioeconomic changes and political stability, are also included in order to highlight additional pressures on policymakers. Stressing these issues emphasizes how environmental factors provided a stimulus to state managers to act.

The Role of the State

Environmental factors alone offer limited insights for understanding policy choice as well as developmental outcomes. However, they do suggest a tendency, a first step towards understanding the context in which export orientation is selected and implemented. In addition, these stimuli tend to be less helpful in accounting for direct policy initiatives at the industry or sector level. In the case of Mexico, for instance, specific features associated with the domestic and international economy during the 1960s led policymakers to search for new export opportunities. These conditions do not, however, help explain why tourism was chosen, or the specific export scheme implemented. As such, the relationship between stimuli and the role of the state is indirect at best. Here it is important to emphasize the policy choices made by state actors as well as the ability to implement them.

First it is relevant to specify what is meant by "the state". Skocpol's (1985:9) neo-Weberian, non-instrumental conception of the state is useful in that it treats the state as a multi-faceted and potentially independent entity *vis-à-vis* civil society. The state may be distinguished from the government in that it includes not only elected officials but also permanent institutions and the elites who staff them. The state is best approached as a complex organization that is both an actor and an arena for politics. Therefore, it is not determined but neither is it solely an autonomous agent. It is conditioned by the interests of societal groups, but ultimately constitutes an independent entity capable of formulating and pursuing its own goals. In such a case, state action must be considered within a larger social context. It need not, however, be the mere product of group or class preferences, but instead may derive from self-generated state interests (Skocpol, 1985). This definition makes it possible to link state attributes and action with developmental outcomes.

Under this general heading of the role of the state, two elements require elaboration: what explains the export choice made and what determines the ability to successfully implement policy. While environmental stimuli constitute a set of exogenously generated incentives facing policymakers, features internal to the state apparatus also influence choices made. Among the most important are the dominant set of ideas found among personnel within state institutions for problem solving. Paying attention to who staffs important bureaucracies and the ideas that are held by key decision makers offers insight into how they identify problems and solutions. In other words, state elites exhibit "embedded orientations" that can be understood by analyzing action or even debates over action (Adler, 1987; Bennett & Sharpe, 1985:43; Sikkink, 1991; Whiting, 1992:55–56). These amount to values or preferred outcomes that policies work toward achieving. Whiting (1992), for instance, classifies liberalism and nationalism as the two dominant sets of ideas held by Mexican state elites regarding direct foreign investment. One may similarly identify embedded orientations held by state managers with respect to favored development outcomes (such as industrialization) and the proper role for the state and market in reaching developmental goals. As Centeno (1994) and others point out, Mexican state actors have exhibited an almost religious belief in a "modernizing rationality" going back at least as far as Porfirio Díaz in the late 1800s. This includes a reliance on learned expertise for solving a wide range of problems. Therefore, policy and sets of policies reflect and are derived from those orientations.

There are several possible objections to this. First, how is change explained? Might not policy be reversed when the environment stays the same? The answer is yes. Innovative policies may result from new ideas alone, but this is rare. More commonly they stem from either changes in the environment, or changes within the staffing of state agencies themselves. In fact the Mexican state saw such a change in the 1980s and 1990s as a new class of economic managers who exhibited different educational backgrounds and embedded orientations captured key policymaking posts and initiated a neoliberal revolution (Centeno, 1994). A second objection is that state response to environmental stimuli may contain elements of more than one strategy. In other words, state managers may pursue a liberal strategy of export orientation in some economic activities while simultaneously engaging in a market-violating export push in others. One possibility is that such a strategy reflects alternate sets of orientations held by policymakers who are placed in charge of different sectors. In fact there are also practical reasons to expect such a mixed strategy at the economy-wide level, especially if policymakers are not wedded to general liberalization. Because of the resources required, an economy-wide export push strategy is unlikely if not impossible. Instead state managers may choose to push certain industrial exports while simultaneously making other industries more market oriented. This amounts to another argument for studying industries. Within individual activities one may examine the specific strategies adopted and identify them as belonging somewhere on the continuum between export adequate (market oriented) and export push.

It is also important to acknowledge that emphasis on exogenous stimuli combined with embedded orientations of state officials is not meant to capture every instance of policy initiation or change. Attempting such an endeavor would be fruitless. Instead the argument is that in order to take a first step toward understanding the timing and method of export promotion selected, attention to environmental stimuli *combined with* the dominant ideas or preferences held by state actors is necessary. In the case of tourism promotion in Mexico, this means attention must be devoted to political and economic conditions prevailing during the 1960s, which served as a stimulus to export promotion. However, equally important is uncovering the embedded orientations held by those charged with managing the economy and especially identifying export opportunities. As is demonstrated below, both factors supported, but did not determine, the export push method selected. Chapters 3 and 4 show that chronic balance of payments problems, import substitution bottlenecks, labor market pressures, and regional development needs constituted incentives for action but did not amount to a crisis. This left a relatively broad set of options available to policymakers.

Within the state apparatus, a growing tendency toward state-directed development also prevailed, especially in the area of planning, financing, and entrepreneurship (Bennett & Sharpe, 1982). This is a product of *both* the social basis of the Mexican state, as Chapter 3 will discuss, and the prevailing ideas or embedded orientations held by elites regarding development. While state actors who staffed the agencies in which the tourism export push originated were more outward oriented and probably more market embracing than counterparts elsewhere in the bureaucracy, their vision for export promotion included a significant role for state intervention, and for manipulation of market signals. The result was an export push that centered on tourism and emphasized planning and a significant state role.

"Choosing" an export promotion program, however, is not enough. Instead the ability to carry out such a project is a product of the public functions selected by managers and their compatibility to existing state structures. In some respects the internal institutional requirements tend to be greater for state-led export pushes than they are for market-reliant export orientation. In the latter case, the state must be able to overcome resistance from societal groups to its economic plan, but this is more an issue of state power or autonomy. An export push may or may not result in similar opposition among societal groups, but additionally it frequently requires more demanding bureaucratic expertise and the ability to command and mobilize resources. These requirements are by no means guaranteed, however. Instead they are contingencies that affect developmental outcomes.

A first step, then, raises the issue of capacity. Evans, Rueschemeyer and Skopcol define capacity as "specific organizational structures the presence (or absence) of which seem critical to the ability of state authorities to undertake given tasks" (1985:351). In other words, does the state have the organizational, technical, and financial means to deal effectively with the developmental tasks at hand? The literature on capacity blurs issues of competence to formulate coherent policy and the ability to penetrate society, or what Shafer (1994) calls absolute versus relative capacity. Geddes places emphasis on capacity as power to penetrate: "… the ability to tax, coerce, shape the incentives facing private actors, and make effective bureaucratic decisions during the course of implementation". This may be compared with her definition two pages later as "expertise in government and the emergence of consensus among political leaders regarding needed changes" (1994:14). Barnett (1992) suggests capacity relates to the means state actors possess in order to govern society. Capacity for purposes here involves both design and implementation of policies but places greater emphasis on the former. It therefore inquires about organizational structure, cohesiveness, and functional expertise. As the discussion below indicates, however, this does not ignore resources and favored policy instruments. According to Sikkink (1991), capacity cannot be treated as an absolute because it depends upon exactly which tasks are taken up by the state; the capacity requirements for states pursuing revolution from above are clearly different than those who take on a night watchman role over the development process. In short, the two must be considered in tandem.

For export promotion, sufficient state capacity is that which meshes the demands of the industry in question with the nature of the roles played by the state (Amsden, 1992:61–66; Evans, 1995). Four state roles in the economy are important for purposes here: regulator, entrepreneur, banker, and motivator. The roles selected by state actors are an open question. They are likely influenced by embedded orientations but also tend to vary to at least some degree by industry. Hence Mexican state managers have historically pursued state rectorship of the economy, but this has taken various forms. For instance, the state has acted primarily as regulator in the banking sector (except when the banks were nationalized between 1982 and 1992) at the same time it acted mainly as entrepreneur in energy. What is crucial is the proper wedding of institutional capacity with the roles selected. Capacity exists when institutions possess the necessary staffing, expertise, and resources with which to successfully carry out the selected roles. This is a necessary, though not sufficient condition for the success of an export push.

For the case at hand, examining capacity begins by identifying and analyzing internal aspects of the state apparatus, including organizational factors, staffing expertise and cohesion, and available resources. This is important both at the general state level, as well as within individual bureaucracies dealing with tourism. Mexico possessed an authoritarian political regime and strong state apparatus during much of the last century. It is characterized by low degrees of pluralism and a relatively insulated institutional structure. Most analysts also consider the Mexican state to contain functionally differentiated and organizationally cohesive bureaucracies (Bailey, 1988; Bennett & Sharpe, 1985:47–50; Whiting, 1992:38–42). The post-revolutionary state has compiled a long record as regulator, producer, banker, and motivator (Bennett & Sharpe, 1978; Hansen, 1974; Whiting, 1992). State dealings with the tourism industry, however, took on a qualitatively different scope after 1968, thereby presenting new challenges to state institutions. The tasks taken on by elites moved from simple promotion and regulation to planning, providing infrastructure, banking, and ultimately production. Therefore, the demands for state capacity grew immensely.

Chapter 4 treats these questions in detail, arguing the state did enjoy sufficient capacity to create an *independent* vision of Mexican tourism, especially at an early stage, as well as the technical expertise and material resources necessary to make the vision reality. One aspect was a shift in institutional arrangements that created a relatively insulated and well-endowed bureaucracy staffed by individuals who held a shared vision of tourism promotion. Most important among these embedded orientations toward tourism were planning and export orientation. The former was quite common in Mexico during the 1960s and 1970s; the latter was much less so. The state also possessed or gained access to necessary policy instruments for implementing the tourism vision. Most important among these were financial and productive policy instruments. Publicly-owned banks and tourism enterprises allowed for state actors to directly implement their plans. In addition, they served to attract private actors into the industry.

Aside from internal state attributes, the second requirement for a successful export push is relational: does the state have the ability to penetrate society in order to put its plan into action? While this raises the question of autonomy, the task here is not to make a conclusive claim about the Mexican case. Debates over state autonomy — both at the theoretical level and in the specific case of Mexico — are long, drawn-out affairs that are not easily settled. State autonomy is a question widely taken up in both neo-Weberian and neo-Marxist literature on political economy (Block, 1987; Carnoy, 1984; Clarke, 1991; Evans, Rueschemeyer & Skocpol, 1985; Haggard, 1990; Lindblom, 1977; Poulantzas, 1978) and also has been dealt with in the Mexican case (Cypher, 1990; Hamilton, 1982; Smith, 1979). The most common question to be answered in capitalist societies is whether the state is autonomous from capital, both domestic and foreign, in taking action. The primary difficulty here is knowing autonomy when we see it. Hypothetically, state policy may be independently formulated and put into action but still be consistent with the interests of dominant societal actors. Grindle (1986:18) argues that when state policy is consistent with the desires of dominant groups, three possibilities exist:

> This may be the result of a weak state that is in fact dominated by
> societal interests, it may be the result of the conviction on the
> part of state elites that these policies are in fact the best (or most
> feasible) means of achieving national development, or it may be
> the result of interactions of bargaining, conflict, and compromise
> between state elites and social classes.

Eric Nordlinger (1981) posits three types of autonomous state: one that acts against societal preferences, one that changes societal preferences, and one that agrees with societal pressures. The first two suggest observable verification of autonomy deriving from conflict, while the third is more difficult. These suggest that empirical verification of autonomy is difficult if not impossible. Instead, as Geddes suggests, autonomy tends to be most commonly inferred and assigned *post hoc* when an unexpected outcome occurs (1994:2–7).

Within the Mexican case two positions tend to stand out. The first is that powerful social groups, most notably domestic business, are subordinate to an independent, relatively autonomous state (Camp, 1989; Smith, 1979), while the second contends that autonomous state action is severely limited due to the structural power of capital (Cockroft, 1983; Hamilton, 1982; Teichman, 1988). No claims are made here that the ability to promote tourism exports constitutes a fundamental test of Mexican state autonomy. Less interesting for the case at hand is the question of whether the state may act against the fundamental interests of capital than the question of whether the state may simply act independently in implementing a conscious and consistent policy of export promotion. In such a case, the concern is with something closer to instrumental autonomy (the ability to act independently of direct class or group pressures) as opposed to structural autonomy (acting against the clear or real interests of dominant groups or classes). The latter, as Hamilton points out, requires an attempt to undermine the existing mode of production in favor of a new one (1982:10–13).

More useful is to consider the relationship between state actors and societal groups to be contingent, depending on the attributes of each considered separately. In addition, the relationship is likely to vary over time and with respect to the policy area. While autonomy is a product of macro-historical forces, industries also matter. The state's relationship to dominant fractions of capital may vary with respect to sector or industry. As Grindle argues, the state may have more room to maneuver in industries where class or group interests are absent or not well organized (1986:15). In addition, some industries — such as steel, automobiles, and oil — come to hold a special place within the national imagery. In short, a more inductive and fluid approach to autonomy is needed. In cases of export push, instrumental autonomy is most likely to occur in just these cases, where industries are new or are not well developed. In other words, the state may very well find great difficulty in changing the trajectory of a leading sector from inward to outward orientation, especially if the activity is dominated by a small number of domestic or foreign capitalists. This was in fact the finding of Bennett and Sharpe (1985) with respect to Mexico's well developed auto industry in the 1970s. In contrast, state action in infant industries will likely be less constrained unless policies directly or indirectly threaten the perceived interests of powerful groups.

If such conditions exist, the question again is what the expected outcome will be. If rational actor assumptions are adopted, as some statist approaches do, might one not expect autonomous state managers to "go into business for themselves"? Haggard (1990), for instance, adopts such language to argue development is primarily a collective action problem: societal groups face organizing difficulties, but once formed they are likely to turn development into a distributional game. Rather than make sacrifices to pursue long-term benefits that are likely to accrue to many, they pursue their own short-term gains at the expense of others. For Haggard, as well as Adler (1987), the answer is a strong state with selfless or nationalistic bureaucrats insulated from such demands, but he does not make clear why rationally motivated public officials should be immune from the same incentives facing private sector actors. Each, acting rationally, would likely seek rents. Ultimately, he contends, international shocks may constrain predatory behavior by public sector officials, but this is unconvincing in that frequently identical or similar shocks lead to different outcomes.

Evans (1992, 1995) offers the most convincing answer. Successful state-led development requires autonomy combined with a sense of "embeddedness" by bureaucrats in the societies they govern. Embeddedness lacks a precise definition but includes a spirit of élan, internal cohesiveness, and a sense of having a stake in society in general and national development in particular. It combines insulation with what Evans calls "intense immersion in the surrounding social structure" (1992:154). In other words, state managers must have a connection with society or groups that gives them a stake in developmental outcomes that varies from staying in power, providing favors for friends, or self enrichment. Industrial transformation will still take place in the absence of embeddedness, but outcomes are likely to be plagued by inefficiencies, predatory behavior, and suboptimal growth patterns.

Two additional points are worthy of mention. One is that autonomy is not static. As industries mature, and especially if they are dynamic, the likelihood is that private interests will form and strengthen, thereby placing increasing limits on state manoeuvrability. The other is that state action is not only contingent on internal capacity or autonomy based on domestic relational factors; it is also conditioned by external factors. The availability of financial resources from external sources, for instance, may expand the domestic manoeuvrability for state actors, perhaps even to the point of confronting societal groups. Ellinson and Gereffi (1990:374–376), for instance, contend that foreign aid to South Korea and Taiwan enhanced state power and autonomy *vis-à-vis* civil society. However, this strategy is seldom without costs and frequently the result is trading international constraints for domestic ones. Barnett (1992:33–34) makes this point over issues of defense and war preparation, but the point also holds for development (O'Hearn, 1990).

With respect to Mexican tourism, what is required is an understanding of the state as well as the nature of the industry and the interests that surrounded it during the initiation of the export push. Chapters 3 and 4 offer such an examination. The former contains a general discussion of the post-Revolutionary Mexican state and situates it as both in its global context and with relationship to key societal groups. It then turns to the contradictions raised by the prevailing development model and the pressures that this puts on state elites. Next, an overview of the history of the tourism industry in Mexico is instructive; it demonstrates that because the industry was in its infancy and interests were not yet well organized, state actors did not directly face powerful and entrenched interests. In addition, because these officials did not pursue tourism at the real or perceived expense of other existing industry, they had greater latitude in pursuing the export push they envisioned. In fact, one of the greatest challenges facing state managers was creating sufficient business interest in the tourism industry. Under these conditions the state is likely to play a central role in shaping the direction of the export push and it did in this case. The content of that direction was related to responses to general macroeconomic conditions and the embedded orientations of policymakers, and to the instruments available for implementing policy.

This does not, however, settle the question of autonomy and its relationship to the industrial transformation of Mexican tourism. With few constraints on state managers, why did they not become more predatory and corrupt? The answer is complex. It relates in part to the nature of the industry, as well as to the specific bureaucracies involved and those who staffed them. As Chapter 4 demonstrates, this latter factor amounted to something akin to embeddedness. In addition, as later chapters will demonstrate, the international tourism industry also placed some constraints on such activity. Ironically, as the closing chapter points out, widespread corruption in Mexico has taken off over the past decade and to be sure tourism has not been immune to the practice. The larger point, however, is that during the initial tourism push corruption was present but predatory behavior was not rampant.

By way of summary, thus far a number of contingencies surrounding export promotion have been posited that range from the circumstances facilitating its initiation to institutional requirements necessary for successful industrial transformation. In the case of Mexican tourism the export push may be traced to two sets of factors: state managers did *not* face a crisis in the period leading up to export promotion and those who undertook the project appeared to share similar embedded orientations. The result was a conscious export push of the tourism industry that involved a significant state role in planning and manipulation of market signals. Next is the question of internal state capabilities (capacity) and relationships with key societal actors (autonomy). The suggestion above is that in the case of Mexican tourism all the prerequisites listed thus far for a successful export push were met: the bureaucracies that carried out its promotion appeared to possess the necessary expertise and policy instruments. In addition, because the industry was in its infancy, the state faced little opposition from powerful societal groups. At least some degree of embeddedness was present, which would suggest that the outcome would not be plagued by predatory behavior.

With all these conditions having been met, the nature of that which needs explaining changes. Why was industrial transformation not even more successful? The answer lies with the final component of the model, the nature of the economic activity itself. Industry characteristics, including the nature of the product, resulting firm assets, and structural consequences placed limitations on both the nature of state action and developmental outcomes. In its simplest terms, the argument is that the state's ability to influence industrial transformation is limited. Of course this is not enough in itself. The section above indicates that internal and relational aspects may account for some of those limitations. In addition, industries themselves shape development outcomes independent of and sometimes despite state action.

Industry Characteristics and Industrial Transformation

Because export promotion involves greater integration with the global economy, features associated with global industries affect local developmental possibilities. More specifically, attributes of individual economic activities influence the local outcomes of export promotion because structures associated with the international industry tend to be reproduced locally. State actors promoting such industries are confronted by these global industrial characteristics. In one sense this is obvious. Public officials attempting to build an internationally competitive computer industry at the local level had better know something about the nature of that industry globally if they want to find a niche. In addition, however, the characteristics of such industries create costs and benefits for state action. Two sets of factors are important here, the first being industry attributes and the resulting structures they create and, second, the degree to which states are willing or able to alter these prevailing patterns of industrial development. Each influences the ultimate transformation outcome of industrial, especially the distribution of material benefits.

Some strands of industrial organization theory have focused upon firm strategy and behavior, and in particular why firms invest abroad. Often it is utilized in studying industries under conditions of less than perfect competition as a means to document and understand performance that is less than Pareto optimal (Bennett & Sharpe, 1985:65–68; Newfarmer, 1985a; Reid, 1987). A central issue is the structure of industries, which results from firm assets and resulting strategies (Bain, 1968; Dunning, 1974; Hymer, 1976; Reid, 1987). Here structure refers to the number of firms within the industry and patterns of ownership and control (Bennett & Sharpe, 1985:66). These structures evolve historically and are produced by the previous actions of firms and market conditions. While firm assets and structural factors are not necessarily seen as determinant of firm behavior, they do shape it significantly and thus are important elements in explaining development outcomes (Bennett & Sharpe, 1985; Newfarmer, 1985a).

Central to this approach is beginning with the specific nature of production and resulting firm assets. These assets are generated by sectorally specific sets of characteristics surrounding the production process, including capital intensity, technological requirements, economies of scale, flexibilities in production, and marketing expertise. Because they are specific to individual economic activities, they tend to produce different industry structures. Hence the capital and technology requirements combined with market demand are such that there are few international producers of commercial aircraft (Golich, 1992) while producers of textiles proliferate (Evans, 1995; Shafer, 1994).

There are two implications for the argument here. First, international industrial structures tend to be reproduced at a domestic level. This is especially the case when the local industry establishes international linkages. The distributional implications are relatively clear. If global industrial structures are characterized by small numbers of leading international firms, local industrial transformation will probably also contain oligopolies, and they will likely contain those same TNCs. As a result a broad distribution of benefits is unlikely. Shafer's ideal types of high/high and low/low sectors are examples of this type of industry or sectoral approach. The former is characterized both internationally and locally by oligopoly due to high capital intensity, economies of scale, and production and asset/factor inflexibility. Although he goes much further than the analysis here for the consequences of industry characteristics, his distinction is useful for understanding the limitations and possibilities for development outcomes (1994:14–15).

Second, the state may attempt to alter this tendency, but the probability of success must be viewed in the context of these same forces. Structures and barriers associated with industrial organization place limits on what is possible for policymakers. More concretely state officials dealing with an oligopolistic structure are faced with choices that range from accepting foreign controlled oligopoly to pushing for some redistribution of benefits. Attempts to reallocate these benefits may take the form of dislodging TNCs in favor of local private or state-owned firms, or through bargaining with them over taxes, employment or other benefits. Such redistributive policies, however, are potentially costly, especially if exit is a viable choice for the firms (Bennett & Sharpe, 1985; Biersteker, 1987; Encarnation, 1989; Grieco, 1984; Moran, 1974; Shapiro, 1994).

If states are willing to attempt to alter that tendency for reproduction, one needs to know more about the costs and benefits they are likely to face. Shafer contends states may attempt redistribute rents from oligopolists, but because they are less effective at managing those rents the costs tend to be too high. He argues that in high/high sectors, which are characterized by high capital and technological demands, collective action is required to effectively manage risk and contain unrestricted competition. TNCs, however, are more effective at this than states. On the other hand, introducing new (usually local) firms also results in damaging unrestrained competition (1994:26–30). This may be true, but the costs and benefits may also vary. An additional variable is the ability of firms to substitute production facilities. This is especially important in mass tourism, where beaches in one destination are easily replaced with those elsewhere. Certainly this is a crucial issue for Mexico, which competes with the Caribbean and Central America for many of its foreign tourists. This raises the related question of strategic sectors. Industries may be deemed strategic for several reasons, ranging from their relative importance to a particular economy to historical or cultural factors to economic or non-economic spillover effects including national defense or patrimony. Banking, in other words, faces a different atmosphere for regulation from that of apparel, because of its widely agreed upon strategic status within an economy. The specific situation also matters. Bananas or cocoa may be treated as strategic in some monocrop-based economies, as might tourism in some other places. Some economic activities are broadly recognized as strategic by nation-states, thereby creating international norms for protection (Clancy, 1998; Golich, 1990).

The larger point is that strategic industries tend to alter costs and benefits for state actors. They may be willing to pay higher costs because the benefits, economic or otherwise, are considered to be worth it. On the other hand, strategic status may lower some costs as well. This is the case in airlines, as demonstrated in Chapter 6. There state actors moved to protect a strategic industry through protection and state ownership. While the economic costs were high, the mainly noneconomic benefits associated with a nationally owned airline industry were considered well worth it. Moreover, the costs were lessened to some degree by a global norm for protectionism which all but ensured that a spiral or retaliatory action would not follow.

With few exceptions (Sinclair, Alizadeh, Atieno & Aonunga, 1992), little effort has been devoted to understanding how the international structure of tourism may affect the nature of the industry within individual countries. This is not a single, well-defined industry but instead is made up of several overlapping sectors ranging from integrated lodging and transport activities to the more decentralized and localized ones such as handicrafts and souvenirs. The focus here is upon two sectors, hotels and airlines, that are especially important at both the international and local level. The two make up the primary branches of the overall sector, account for the bulk of tourist expenditure, and are the most integrated at the international level (Britton, 1982; Dunning & McQueen, 1982).

If the model is accurate, one should expect the industrial attributes of each international subsector to influence significantly the nature of change within Mexican tourism. Moreover, because the industry is made up of overlapping sectors that have their own unique production processes, one should expect variation *among* the different components that make up the tourism product. If oligopoly characterizes the structure of a particular global sector, a similar structure is likely to evolve domestically. Under such conditions state managers may attempt to redistribute these benefits, depending on their calculus of the costs and benefits associated with such a strategy. Because it alters those effects, one factor that will influence state action will be the strategic nature of sectors. Strategic industries invite state attempts to alter local industrial patterns.

Conclusion: Arguments and Objections

An analytical framework for understanding the logic behind developmental outcomes that result from export promotion is presented above. It examines three sets of factors that shape industrial transformation: the environment that leads to export promotion, features associated with state actors and institutions, and industrial characteristics. Constructing such a model helps approach the case of Mexican tourism in a theoretically informed manner in that it points to an underlying logic that accounts for the specific path that the export-led industry has taken. In addition three arguments derive from the model.

First, a state-initiated export push, defined above as market violating, may successfully promote growth given certain structural conditions associated with the state apparatus and the economic activity. As subsequent chapters will show, with respect to Mexican tourism these conditions were met. As a result, state-led export promotion produced a better growth record than almost certainly would have been achieved through reliance on the market alone. This argument runs directly counter to neoclassical expectations (Balassa *et al.*, 1986; World Bank, 1993) and supports some statist or institutionalist claims (Amsden, 1992, 1989; Evans, 1995, 1992; Haggard, 1990; Haggard & Moon, 1989). Next, state-led, export-led growth strategies *may* aid in enhancing local benefits associated with industrial transformation. Again, as Chapters 5 and 6 will demonstrate, Mexican tourism was not captured by foreigners as some strands of dependency theory might predict. Ultimately, however, the fact that those who own and control the Mexican industry are made up of a fraction of international and domestic capital suggests that the benefits associated with tourism exports accrue mainly to the few. Because this outcome is traced primarily to the international industrial organization of the industry, which itself derives from industry-specific characteristics, this suggests that broad-based developmental prospects for tourism in Third World development are limited. This is not to suggest that tourism should be completely scrapped there or elsewhere in the Third World. Only careful consideration of alternative developmental possibilities can determine if this should be the case.

Finally, while taking the state's role in formulating and implementing development strategies into account rightly corrects against purely systemic or societal explanations for developmental outcomes, there is a danger in attributing excessive explanatory power to state policy. Instead greater attention needs to be paid to the nature of industries and their independent role in shaping developmental outcomes. That is, while Grindle and Thomas argue, "the primacy of policy as the basis for encouraging and sustaining economic growth and social welfare had come to be widely accepted by those concerned about promoting development" (1991:xi), it is not the only determinant. Instead the return of the state need not and must not completely negate other explanatory factors. The framework provided here can be used to analyze both state action and market and industry determinants of developmental outcomes, not only for tourism in Mexico and elsewhere in the former Third World, but also for other export-led growth strategies.

The next four chapters utilize the framework presented here to examine the course of tourism development in Mexico over a 25-year period. The task is not to "prove" the model but instead to present a theoretically informed treatment of tourism development in Mexico and in turn ask how the facts of the case fit with the framework provided here (Barnett, 1992:50; Shafer, 1994:15). Chapter 3 provides a broad historical background for analyzing the stimulus surrounding the choice for tourism as export promotion. Chapter 4 details the choice made and resulting specific nature of the export push. It also highlights the embedded orientations found within bureaucracies and discusses the roles played by the state, the resources it mobilized, and the policy instruments it utilized in pushing tourism exports. Broad outcomes of the push are also provided. Chapters 5 and 6 shift to an analysis of the subsequent developmental outcomes at a sectoral level by looking at growth and distribution within subindustries. Chapter 5 examines the international and domestic hotel industry in detail while Chapter 6 provides a similar treatment of airlines. Finally, Chapter 7 concludes with an update on Mexican tourism under liberalization in the 1990s and reassesses the analytical model.

Chapter 3

The Stimulus: Import Substitution and the Global Tourism Boom

In the mid-1960s the *Banco de México*, the nation's central bank, undertook a thorough study aimed at identifying and expanding export opportunities. Three years later it released its final report, which called for providing tourism activities for foreigners as one central answer to the country's trade and development needs. The plan called for the establishment of a series of new resorts, the refurbishing of existing ones, and the aggressive marketing of the country to potential tourists from abroad. Over the next 25 years, five presidents, ranging from free spending populists to fiscal conservatives, dutifully and somewhat unwaveringly carried out the original program called for by the central bank. If anything, the initial plans grew larger in scope. Mexico, and especially its government, came to embrace tourism, and in particular the *foreign* tourist.

In order to understand why this has been the case, it is first necessary to integrate the decision to promote exports, specifically tourism, into a broader context. The framework laid out in the previous chapter indicates that one needs to examine the larger macroeconomic and political environment surrounding the initiation of export promotion in order to more fully understand its timing and content. More specifically, the export push that has characterized tourism policy in Mexico would be more likely to come about under conditions that demanded policy innovation but did not constitute an economic or political crisis. This chapter highlights the setting under which the decision to promote exports, specifically tourism, took place. It identifies the specific problems policymakers faced and the goals they had in mind when choosing tourism and state-led growth.

The chapter highlights three areas. One, it summarizes the historical evolution of the Mexican state and its political economy during the post-revolutionary 20th century, paying special attention to development models and their impact on the external sector. This closes by discussing macroeconomic, social, and political pressures that either appeared or were heightened in the

1960s, leading up to the decision to promote tourism. Two, the chapter provides a supplement of sorts to this treatment by examining the nature of global and local tourism. Further, it documents the growth of global tourism during this period and discusses its growing popularity among influential policymakers. Three, it looks at tourism in Mexico through 1968, when the *Banco de México* made public its completed study. Together the chapter helps answer two central questions: why export promotion in the late 1960s, and why tourism?

Mexican Development, 1930–1970

Although debate continues on the nature of the post-revolutionary Mexican state, the dominant position holds that the revolution did not eclipse the previous social basis of the state (Cockroft, 1983; Hamilton, 1982; Meyer, 1977; Saragoza, 1988). Although the revolution mobilized the popular classes, their institutional incorporation into the dominant political party, the Institutional Revolutionary Party (PRI) in the 1930s, ultimately served as a means of control and effectively modernized the Mexican state. The PRI, which was founded in an earlier incarnation in 1929, has ruled the Mexican political landscape ever since, although it has resorted to electoral fraud, mass patronage, and selective repression in doing so. The historic basis of the ruling party has been the peasantry, labor, and the popular classes, although organization of those sectors has taken on a corporatist nature. As a result, the interests of these subordinate groups have not been fully represented by state action. Also notably absent have been leading business groups, and in fact much of this century has been marked by alternating periods of distrust and open enmity between government and private sector (Maxfield & Anzalua, 1987). With low levels of political democracy combined with little public support from domestic capitalists, state officials traditionally relied on providing material benefits to citizens in order to maintain political stability and support.

Within such a political climate Mexico produced by the 1960s, by most traditional measures, an enviable development record. The well-known "Mexican Miracle" combined 30 years of political stability with steady economic expansion. Fueled by an import substitution-based development model, the country registered annual average growth rates of roughly six percent over this 30-year period. In fact the PRI certainly owed a significant portion of its longevity to the developmental miracle, at least until 1982 when the debt crisis struck. During this time the structure of the Mexican economy — and society — was fundamentally transformed, shifting from an agricultural to an industrial base and from a rural nation to an urban one. Yet it is also important to note that while a middle sector was largely created under this miracle (although significant portions of it emigrated from Spain during the late 1930s and 1940s), this period of economic expansion provided very few benefits for many Mexicans, especially those living in the rural sector. The primary question to be addressed here is whether evolving pressures on the existing development model amounted to a crisis that would favor a market-oriented response or instead were a lower level set of stimuli that would allow for an export push.

The exact origin of import substitution policies continues to be debated. For some (Glade, 1963; Haber, 1989; Reynolds, 1970), the revolution served as a watershed that distinguished 20th century economic policy from previous models. According to this view, Mexican efforts at industrialization in the 1940s and beyond were intimately linked to revolutionary nationalism. Others contend conscious plans for industrialization began with the Depression in 1929 (Gereffi, 1990; Villareal, 1990). In fact it does appear that import substitution policies were adopted on a piecemeal basis when, in response to the disruption associated with the external shocks of world depression and then World War II, the state provided incentives in order to encourage domestically-produced manufactures. Although import substitution was consistent with the ideology of revolutionary nationalism unique to Mexico at the time, similar measures were taken throughout Latin American during this period and came largely as a response to the crisis these shocks produced for primary exports (Baer, 1972).

The policies included providing tax exemptions, the channeling of credit toward industry through the state development bank, NAFINSA, and protection through the introduction of import controls and higher tariffs. Many were embodied in a 1941 law on manufactures and an import licensing system introduced in 1944 but not fully implemented until 1950 (Hansen, 1971; Izquierdo, 1964; Mosk, 1954). Licenses, which affected just one percent of imports in 1947, covered more than 60 percent two decades later (Solís, 1971). The policies had their intended effect, with manufacturing growth outpacing the 6.7 percent annual rate of GDP growth between 1940 and 1950 (Hansen, 1971:42–43). Import substitution accelerated and entered a vertical or deepening phase that emphasized production of consumer durables, intermediate, and some capital goods from 1954–1970. This era was commonly referred to as "Stabilizing Development" due to the achievement of high growth rates with low inflation (Looney, 1978:15).

While the state was still active during this period, Stabilizing Development largely benefited the private sector (Hellman, 1983; Solís, 1981). Mexico was unique in Latin America in that it pursued import substitution in a manner that produced rapid industrialization, sustained growth, and low inflation. Prices had risen by an annual rate of more than eleven percent between 1940 and 1950, corresponding to the early stages of industrialization, and just under ten percent in the period 1951–1955. After devaluation in 1954, however, state spending was reduced and prices stabilized over the next decade. The annual rate of inflation between 1950 and 1960 was 6.4 percent and between 1960 and 1970 fell to 2.4 percent (Graham, 1982). Comparative figures show that Mexico experienced considerably greater price stability during this period than other Latin American countries pursuing import substitution (Kaufman, 1990:123).

As vertical stages of import substitution intensified and production became more dependent on technology and capital inputs, the role of the state began to change. While still playing an important role in financing development, the public share of gross fixed capital formation began to fall (Hansen, 1971:43), and an increasing portion was made up of direct foreign investment. State rectorship of the economy continued, but more in the form of management, first through encouraging "Mexicanization" or domestic ownership and control of strategic industries during a period when direct foreign investment in manufacturing accelerated rapidly (Evans & Gereffi, 1982; Whiting, 1992), and second through expanding state ownership of firms. Most often this was the case in areas where private initiative was lacking, or in strategic industries.

As Table 3.1 demonstrates, the results over the long term were impressive. Economic growth averaged 6.5 percent from 1950 to 1970 and industry grew even faster. By 1970 the industrial portion of GDP reached 34.3 percent, with manufacturing accounting for 22.6 percent. The share of the labor force working in manufacturing by 1970, 18.6 percent, was almost double that of two decades earlier. While nearly two-thirds of Mexicans had worked in the agricultural sector in 1940, that figure had dropped to less than 40 percent by 1970 (Graham, 1982). This was due to a number of factors including land scarcity, population growth, industrialization, and modernization of agriculture. The significance of this change is that, aside from the massive demographic shift itself, the transformation likely pleased policymakers who viewed the primary sector as "unproductive".

By that time the structure of industry had also changed. Vertical import substitution produced a relatively diversified industrial sector. Only about half of industrial value added was in the traditional consumer non-durables area, while the remainder came in the area of intermediate and capital goods (Graham, 1982). In other words, by 1970 Mexico had become a major producer of iron and steel, finished automobiles, chemicals, and electrical and transport equipment. The import share of all manufactured goods, which had been at 48 percent in 1939, fell to 21 percent in 1970. Mexico's overall dependence on trade also declined during this period, and among imports, more than 80 percent of the total constituted capital goods (Looney, 1978; Villareal, 1990).

Table 3.1: Mexican GDP and Industrial Output Growth (1950–1970; annual percent change).

Period	GDP	Industry
1950–1955	5.6	5.6
1955–1960	5.7	7.3
1960–1965	7.1	8.6
1965–1970	6.9	8.9

Source: Graham (1982).

Despite this record policymakers were, by the 1960s, faced with several emerging problems. Most pressing, Mexico suffered from chronic but worsening balance of payments problems, the so-called Achilles heel of ISI. Earlier short-term crises had led to devaluations of the peso in 1948–1949 and 1954, but after the mid-1950s the rate of growth in exports consistently lagged behind that of imports, producing persistent trade and current account deficits. The peso once again became overvalued, making Mexican exports more expensive on the international market and in effect subsidizing imports. By the 1960s the problem became more pressing. Reynolds (1978) estimates the peso was overvalued against the dollar by 9.5 percent in 1960 and 18.7 percent in 1965. In addition the problem had become structural in nature: Mexico ran a trade deficit every year from 1955 through 1982, and as Table 3.2 demonstrates, by the 1960s the deficit became significant, averaging roughly three percent of GDP annually.

In addition to external account pressures there was the growing problem of creating and targeting jobs. A young population and high birth rates meant a huge demand for jobs for new entrants into the workforce. This became more difficult to accomplish as the development model became more capital intensive. Moreover, the urban bias of ISI combined with increasingly mechanized and capital intensive agriculture produced rapid demographic change. Migration from the countryside led to overcrowding in cities and by the early 1960s state officials had become increasingly committed to enhancing regional development schemes. This problem was exacerbated by the United States' unilateral cancellation of the *bracero* program in 1964, which had since 1942 legally admitted Mexican migrants to work in the United States.

On the macroeconomic front, government outlays gradually began to exceed revenues, and the resulting budget deficit expanded significantly during the 1960s. By 1970 the deficit amounted to about three percent of GDP (Looney

Table 3.2: Mexico, Selected Economic Indicators (1955–1970).

Year	GDP Growth %	Merchandise Trade Acct. Millions $	Current Account Millions $	Current Acct. Deficit/GDP %
1955	8.7	−84.8	149.0	2.1
1956	6.2	−233.8	−52.2	−
1957	7.7	−415.2	−220.4	−
1958	5.5	−377.1	−204.6	−
1959	2.9	−259.6	−48.5	−
1960	7.9	−409.0	−324.0	2.6
1961	3.4	−299.0	−227.5	−
1962	4.9	−202.4	−171.3	−
1963	6.3	−254.0	−217.1	−
1964	10.0	−422.0	−413.3	−
1965	5.4	−414.0	−403.0	3.1
1966	7.0	−406.0	−392.0	−
1967	7.1	−596.0	−628.0	−
1968	10.4	−702.0	−744.0	−
1969	8.2	−624.0	−592.0	−
1970	8.5	−898.0	−1,068.0	3.3

Source: NAFINSA (various issues); IMF (various issues).

1978:18–19). Making the problem more difficult was that while the ratio of trade to GDP was falling, the government depended upon taxes derived from trade for roughly one-quarter of all federal revenues (Izquierdo, 1964; NAFINSA various issues). The twin deficits on the budget and current account were the most serious problems facing the Mexican government by 1970. Whether they signified the so-called exhaustion of the import substitution industrialization model at this point remains in question, but they were deemed serious enough during the 1960s that the state began investigating alternatives that would reduce the dual pressures.

In addition to these problems, deeper social and political challenges were emerging. For instance, it was becoming clear that the bulk of the costs of industrialization were being imposed upon the lower classes (Hansen, 1971; Hellman, 1983; Reynolds, 1978) while domestic and foreign capitalists gained most of the fruits of the development "miracle". The capital-intensive vertical phase of ISI exacerbated this problem and resulted in one of the most unequal distributions of income in Latin America, and for that matter the world (Gereffi, 1990; Hellman, 1983; Solís, 1971). Within the middle and popular sectors, pressure for employment, wage increases, and land redistribution increased (Fitzgerald, 1985; Reynolds, 1978). As these demands went unmet, pressure also increased against the authoritarian political regime. Ultimately these led to independent union and peasantry movements, including strikes by middle class workers such as medical workers in 1964 and 1965. In addition various guerrilla movements cropped up, and the overall social unrest and protest culminated in the government's massacre of students and other protesters in Tlatelolco plaza in October, 1968. Business groups also bristled at the economic difficulties and began to withhold investments, despite the government's openly pro-business stands during the latter half of the 1960s.

Essentially by the 1960s the political constituency behind the ISI development project had begun to change. As Hamilton (1986) suggests, this in part was a result of contradictions within the model itself, or what she argues was really two models. The first, nationalist in origin, can be traced to the aftermath of the revolution and the presidency of Lázaro Cárdenas (1934–1940). Its premise was to limit the exposure of the national economy to the international market, and to pursue policies that would benefit peasants, workers, and the popular sector. The second, more internationalist and capitalist in nature, originated in the 1940s and accelerated with vertical ISI in the two subsequent decades. It promoted large-scale economic projects that often required foreign capital, inputs or technology. Agriculture became more mechanized, capital intensive, and technology driven, with production increasingly destined for the international market. Within industry, the state entered into partnerships with, and subsidized local and foreign capital along the lines of Evans' (1979) "triple alliance". Despite continuing nationalist rhetoric, the import substitution industrialization development model amounted to what Hellman (1983) refers to as trickle down economics.

Mexico's adoption of export diversification at this point can only be understood within this larger context of Mexico's political economy. Import substitution had produced significant growth and industrialization but by the 1960s economic and social pressures were mounting. With industrial exports uncompetitive on world markets, state officials were faced with finding an alternative or at least supplementary strategy that was economically viable, would help alleviate the current account deficit, and would create regional employment. While the import substitution model could have been abandoned at this point, in retrospect such a scenario was highly unlikely. Mexico by no means faced an economic crisis, and powerful forces — many within the state apparatus itself — had much to lose if the prevailing development model was reversed. Instead economic and social conditions allowed for tinkering in the form of continued ISI deepening supplemented with targeted export promotion.

Growth and Internationalization through Tourism

While attention to the emerging political economy of Mexico aids in explaining the short-term stimuli that favored the adoption of export orientation, and perhaps even the form of export promotion chosen, it tells little about why state managers chose tourism. This section and the next demonstrate just why this industry appeared so attractive, as it was already booming at the global level, registering annual average growth rates in arrivals of 10.6 percent and 12.6 percent in receipts during the 1950s (WTO, 1999). Yet up through the mid-1960s the activity had been somewhat neglected within Mexico. Thus, the opportunity to exploit its tourism potential was clear, particularly as international development officials and agencies had jumped on the bandwagon, arguing that tourism could enhance development in the Third World.

International tourism is largely a 20th century phenomenon. In fact, most of its growth and dynamism has taken place only after World War II. Statistics on international tourism reflect this trend: figures during the first half of the century lack reliability and are difficult to compare with later numbers. Part of the problem lies with specifying international tourism. Currently accepted practice is to define international tourists as persons visiting another country outside of their permanent residence for more than 24 hours, regardless of the motive for travel. This definition was accepted by the World Bank and International Union of Official Travel Organizations (IUOTO), the precursor to the current World Tourism Organization (WTO), the UN body that deals with tourism issues. Those traveling to another country and staying for less than 24 hours are considered excursionists. These classifications were first agreed upon at a UN conference on tourism held in Rome in 1963. Business travelers are included in international tourism statistics, and by some estimates they make up 40–45 percent of the total. Those traveling on government business make up an additional five percent (Bull, 1991:18).

Much of the growth of international tourism has taken place within the past 40–50 years. At least four factors contributed to this growth. First, post-war economic prosperity in First World countries created higher standards of living and more free time, making tourism available to a wider segment of societies. Second, technological advances, especially in transportation, made traveling greater distances easier. Air travel in particular became more widely available after World War II and especially after the invention of the jet engine. The end of the war also created excess supply of airplanes. International and domestic air travel on scheduled US airlines, for example, increased nearly tenfold between 1940 and 1950 (O'Connor, 1989). In 1950 an average passenger airplane held about 70 passengers and traveled at a top speed of 270 miles per hour. When the first commercial jet airplane was put into service in 1958, both passenger capacity and top speed approximately doubled (Jiménez Martínez 1990). Third, the war itself exposed many to foreign lands. This inspired nostalgic tourism by former soldiers, especially among American G.I.s who returned to Europe (Turner & Ash, 1975:95). Fourth, government restrictions on travel and foreign exchange were gradually lifted following the war (World Bank, 1972:3–5).

As Table 3.3 demonstrates, these factors contributed to rapid growth in international arrivals and receipts. In 1950, international tourist arrivals totaled 25.3 million. A decade later the figure had more than doubled. Between 1950 and 1970 arrivals grew at more than ten percent annually, and receipts (which exclude transportation expenditures) increased at even faster rates (World Bank, 1972:3). Table 3.4 shows the regional distribution of tourist arrivals between 1950 and 1970. Europe remained the top international destination, with its share of arrivals peaking

Table 3.3: Trends in International Tourism (1950–1970).

Year	Passenger Arrivals (Thousands)	Annual Change (%)	Receipts (US $ Millions)	Annual Change (%)
1950	25,282	–	2100	–
1960	69,296	174.1[a]	6867	227.0[a]
1961	75,281	8.6	7284	6.1
1962	81,329	8.0	8029	10.2
1963	89,999	10.7	8887	10.7
1964	104,506	16.1	10,073	13.4
1965	112,729	7.9	11,604	15.2
1966	119,797	6.3	13,340	15.0
1967	129,529	8.1	14,458	8.4
1968	130,899	1.1	14,990	3.7
1969	143,140	9.4	16,800	12.1
1970	159,690	11.6	17,900	6.6

[a]Cumulative percentage increase over the decade.
Source: WTO (1991).

Table 3.4: Regional Share of International Tourist Arrivals (Percentage).

Region	1950	1960	1970
Europe	66.6	72.7	70.8
Americas	29.6	24.1	23.0
East Asia/Pacific	0.8	1.0	3.0
Africa	2.1	1.1	1.5
Middle East	0.8	0.9	1.2
South Asia	0.2	0.3	0.6

Note: Columns do not necessarily add up to 100 due to rounding.
Source: WTO (various years).

in 1970. The Americas ranked second as a region, although both arrivals and receipts to the region diminished during this period before stabilizing after 1970.

Data on the growth of tourism to the Third World at this point are less precise, but available evidence indicates that developing country share of arrivals and especially receipts increased. One estimate suggests that from 1960 to 1968 developing countries increased their share of world arrivals marginally, from 7.2 percent to 8.0 percent of the total, but receipts grew from 17.6 percent to 20.0 percent (UNCTAD 1973:6). Turner (1976) reports that between 1965 and 1971 Third World nations increased their share of the market from 16.6 percent to 19.2 percent. This appeared to be the case in the Western Hemisphere, at least over a longer time period. According to WTO data reported in Harrison (1992:5) the Caribbean share of arrivals in the Americas increased from 6.7 percent to 10.7 percent during this period while the increase to Central America was from 5.2 percent to 8.0 percent.

Much of this growth was market driven. With few exceptions government action in much of the world remained primarily permissive in nature until the late 1960s. This was especially the case in Third World countries, where few significant initiatives toward tourism were undertaken at the state level. For instance, Tancer's (1975) summary of government policies toward tourism in the Western Hemisphere uncovers few explicit actions toward promoting tourism before 1968. Instead regional growth was most often the product of market forces alone.

Rather than an economic focus, international tourism became embroiled in larger global politics during this period. As the Cold War heightened, official Western discourse elevated international tourism to a symbol representing freedom of movement, leisure, and prosperity, all rights and privileges associated with democratic-capitalist societies (Jiménez Martínez, 1990; Moreno Toscano, 1971). Meanwhile, within the Soviet bloc great attention went to emphasizing the volume of tourists serviced as part of state attention to workers' leisure time and social tourism. In the West, emphasis during this period was also placed on the idea that this movement, through facilitating contact among different peoples, would serve to increase understanding and ultimately result in peace (the motto of the World Tourism Organization continues to be "Tourism: Passport to Peace").

In the 1960s the economic promise of this industry combined with this political rationale to encourage governments and international organizations to promote international tourism, or at least discourage barriers. The United Nations held an initial Conference on Tourism and International Travel in Rome in 1963, where it declared tourism to be "… a basic and desirable human activity, meriting the praise and support of all peoples and all governments …" (quoted in OAS, 1979:1). Tourism was also promoted as a facilitator of understanding and conciliation among peoples at the 1962 Interamerican Congress of Tourism, which resolved to recommend that its development be integrated into the Alliance for Progress.

As the 1960s progressed, tourism exports began to be viewed within the Third World as an important element in the balance of payments, a cheap source of jobs and of government revenue. Moreover, they were seen to provide a more stable alternative than traditional agricultural and mining exports, which were susceptible to significant market fluctuations, and of other products which were subject to protectionism (Gray, 1970). Policymakers in wealthy countries also zeroed in on tourism as a development tool for the Third World. Intergovernmental organizations also jumped on the bandwagon. The United Nations embraced international tourism, not only for its alleged contribution to greater international understanding, but also declaring that, "tourism may contribute, and actually does contribute vitally to the economic growth of developing countries" (UN, 1963, quoted in Lanfant, 1980:15). A similar position was taken by the World Bank group, which began making loans to developing countries for tourism related developments in the late 1960s (UNCTAD, 1973; World Bank, 1972). By the end of the decade, a broad consensus among Third World elites, First World governments, international agencies, and the private sector began to emerge in favor of global tourism and especially tourism in developing countries.

Pre-1970 Tourism in Mexico

The general profile of early 20th century tourism growth in Mexico for the most part mirrored world trends. Arrivals and receipts, for instance, accelerated rapidly after World War II, making the industry a leading source of foreign exchange. Mexico clearly benefited from close proximity to the world's largest tourism market, the United States. State action toward the industry in Mexico before the late 1960s, however, was uneven and nonsystematic. While some variation can be found in the attitudes and actions that post-revolutionary presidential administrations undertook, the overarching trend was a reliance on market forces to spearhead the development of tourism. It was not completely ignored, but was left largely to its own devices and at best was seen as an instrument toward other developmental goals.

For much of the first half of the 20th century, tourism was scarcely recognized as an area of economic activity in Mexico. According to one estimate (Ramírez Blanco, 1983:36) 8,000 foreign tourists visited Mexico in 1920, almost all coming from the United States. The state offered only limited recognition of tourism as an economic activity until after World War II. Most mention of tourism was confined to population or migration laws. The Mexican government first recognized foreign tourists in the Federal Migration Law in 1926 (Ramírez Blanco, 1983). It first created a joint public–private sector commission to study and promote tourism in 1929. A year later the study group was transformed into the National Tourism Commission (Caballero, 1984; Torruco Marqués, 1988). The first known government study of the economic impact of tourism on the economy was undertaken by the Mexican central bank in 1942. Though various declarations did create government departments, councils and commissions charged with promoting incoming tourism, they enjoyed little prestige or financing. Each was soon replaced by a successor and each new public tourism body found itself being juggled from one cabinet-level ministry to another.

Despite this the number of arrivals grew from 30,000 to 139,000 between 1930 and 1939 (SECTUR, 1991:59). Tourists, however, continued to be concentrated in the largest cities, especially Mexico City, and along the border. Most travel to the latter was by automobile and originated in California, Texas, and other parts of the US southwest (Jud, 1974:26). Additional tourism was largely confined to already established rail and sea links. The factor that most limited travel elsewhere in the country was a lack of infrastructure, especially roads and airports. In addition, the aggregate numbers were somewhat misleading. Of the 1930 total, more than 60 percent were made up of Mexican nationals living abroad. Because the figures also did not distinguish among visitors staying overnight and those making day trips, tourists were also over-counted (Jiménez Martínez, 1990:23). Most non-Mexican nationals who made up this number were well-to-do Americans interested in cultural tourism.

After World War II Mexico steadily grew as an international destination. More than any other factor, Mexico's close proximity to the United States aided growth immensely and between 1945 and 1950 foreign arrivals to Mexico more than doubled, from 164,000 to 401,700 (SECTUR, 1991:59). The totals were aided in part by the rapid development of Acapulco as an international coastal resort destination after roads linked the town to cities in the interior. International discovery, first by the wealthy and then the middle class in the United States, produced explosive population growth in the town. The village quickly became a city, the result of what Turner and Ash call "the first international resort to have depended primarily on air-borne tourists" (1975:94). The post-war boom in tourism drew some government attention, although tourism continued to remain a low priority and was left to low-level commissions. The bulk of public investment was channeled toward the manufacturing sector to aid ISI.

A significant shift of attitude toward tourism, however, took place within the administration of Miguel Alemán Velasco (1946–1952). Specifically, Alemán recognized the potential for further growth in Acapulco and under his administration transportation links to the city were improved. A new international airport was built and construction of a direct Mexico City–Acapulco highway was initiated (Ramírez Saiz, 1989:140). In part due to the rapid development of Acapulco during his presidency, Alemán has come to be known as the "father of Mexican tourism". Some (Cockroft, 1983:152; Shacochis, 1989:44–45) have argued, however, that rather than simply recognizing the potential for tourism growth in the country, Alemán had a personal stake in its development through his prior purchase of tracts of land in Acapulco. He was also among the first Mexican presidents to identify tourism as a viable source of foreign exchange earnings. As early as the 1946 presidential campaign he singled out earnings from tourism exports as a source for financing imported intermediate and capital inputs for industrial development (Jiménez Martínez, 1990:45–46). Alemán also oversaw the establishment of the National Tourism Commission (CONATUR), a joint public–private council charged with the promotion, study, and consultation of tourism in 1947. In 1949 the first Federal Tourism Law was enacted, mandating further promotional and regulatory activities (Caballero, 1984; Torruco Marqués, 1988).

After 1950 statistics on tourism to Mexico became more reliable, although measurement problems remained. Government methods failed to distinguish between international tourists who spent at least one night in the country and excursionists or those taking day trips for shopping, business, or other purposes. Data on expenditures were similarly inflated due to definitional and measurement problems. Many border transactions were recorded as tourism imports and exports when in fact they constituted merchandise trade (Izquierdo, 1964; Jud, 1974). As a result, most studies during that time tended to significantly overstate the importance of tourism to the Mexican economy. UNCTAD (1973:8), for instance, reported that tourism receipts accounted for 46 percent of all Mexican exports in 1968. The true figure was more likely around ten to twelve percent.

Table 3.5, which corrects for these problems, summarizes the growth of international arrivals and receipts to Mexico between 1950 and 1970. It shows arrivals growing by an annual rate of 9.1 percent during this period, and receipts increasing by 7.3 percent annually. Most of this growth took place in the 1960s when receipts grew by more than eleven percent. From 1950 to 1970 this export grew twice as fast as all Mexican exports and constituted roughly four percent of gross domestic product (Jud 1974:21, 29).

Table 3.5: Tourism Growth in Mexico, 1950–1970.

Year	International Arrivals[a]	International Receipts[b]	Tourism exports/ Total Exports (%)
1950	402	111	13.4
1951	443	111	–
1952	460	115	–
1953	439	109	–
1954	526	86	–
1955	564	118	–
1956	626	134	–
1957	662	129	–
1958	697	134	–
1959	746	145	–
1960	761	139	7.5
1961	803	142	–
1962	941	161	–
1963	1,058	190	–
1964	1,210	215	–
1965	1,350	245	12.3
1966	1,499	268	–
1967	1,629	292	–
1968	1,879	337	–
1969	2,065	373	–
1970	2,250	415	14.1

[a]Thousands of people, not including excursionists.
[b]Millions of US dollars, not including border transactions.
Source: SECTUR (1992:59, 100); IMF (various issues).

The majority of this growth, however, remained market driven. Mexico continued to benefit from its proximity to the large US market, with more than 80 percent of international arrivals originating there. Increasingly, however, the type of tourists that the country attracted became a concern. Border tourism, as it was defined at the time, constituted almost 60 percent of the total (Jud, 1974:21). Short stays contributed to relatively low spending per visit, and often foreign exchange leakage was high. In addition, the reputation of some border areas as centers of vice and smuggling intensified. Of the remaining 40 percent who traveled to the interior of Mexico, most went to Mexico City and to a lesser degree Acapulco.

Between 1950 and 1970 various Mexican administrations had a somewhat mixed and uncertain attitude toward inbound tourism. In public speeches, President Ruiz Cortines (1952–1958) reiterated the importance of tourism for providing valuable foreign exchange (Jiménez Martínez, 1990:48). But he also cautioned against the negative social impact that international tourism might create. Ruiz Cortines responded to the problems of vice by urging Mexicans in speeches to maintain the traditional cultural identity or "lo Mexicano" among the emerging tourism zones. He also sought to promote more domestic tourism in these areas (Jiménez Martínez, 1990:48–50).

Despite mixed feelings regarding the social consequences of tourism development, it was Ruiz Cortines who oversaw the creation of the Tourism Guarantee and Promotion Fund (FOGATUR) in 1956. This served as both an institution of tourism promotion and a trust fund within the national development bank (NAFINSA). Its primary task was to underwrite private loans for tourism-related projects. The significance of the agency was perhaps most important, however, at a symbolic level. The initial endowment of the fund was quite small (Jud, 1974:25–26; Ramírez Blanco, 1983:44), but its establishment marked an important turning point. For the first time the government created a federal entity to actively channel resources into tourism development. Its creation also suggests that state officials recognized a need to stimulate supply of tourism facilities, especially because the industry failed to keep up with the growth rate of world arrivals during the 1950s. In fact, Mexico's share of international tourists fell from 1.6 percent in 1950 to 1.1 percent in 1960 (Jiménez Martínez, 1990:59). This performance was at least partially due to the continued lack of hotel space in the country. The number of hotels totaled just over 3,000 in 1960 with about 87,000 rooms, but approximately one-third of the total was in the Mexico City area (Jiménez Martínez, 1990:55).

Adolfo López Mateos (1958–1964) and Gustavo Díaz Ordaz (1964–1970) also wavered somewhat in their enthusiasm for tourism. López Mateos declared in a speech in Rio de Janeiro in 1960 that aside from the economic importance of tourism it aided "better understanding between peoples". After the Department of Tourism produced a report entitled *Tourism as a Means of Human Communication* in 1967, Díaz Ordaz referred to it as a work that underscored the fact that "tourism is, over all else, a means for better understanding and friendship between men." (Jiménez Martínez, 1990:86–87). Yet both were also hesitant to promote tourism more strongly. This resulted mainly from continued social problems associated with border tourism. Also significant was the conspicuous absence of coastal tourism promotion under these presidents. Despite these concerns, state action toward this industry became more visible. The Department of Tourism, which had been a small sub-agency within the Department of the Interior, was raised to the level of an autonomous government body in December of 1958, a designation roughly the equivalent to an agency in the United States (Ramírez Blanco, 1981). In December of 1961 the federal government authorized the creation of the National Tourism Council, which was charged with consulting and advising federal authorities involved in tourism (Caballero, 1984; Ramírez Blanco, 1981). The council, headed by former president Miguel Alemán and made up of individuals from both the public and private sectors, came to enjoy considerable prestige, in part due to Alemán, but also because of its mandate as a consultative and advisory body. In addition, by 1964 Mexico's Department of Tourism had established 15 promotional offices abroad, including twelve in the United States, two in Canada and one in Argentina. Significantly, Mexico secured the 1968 Olympic games and the 1970 soccer World Cup.

Despite this recognition of the importance of tourism, government initiatives remained mostly limited to promotion along with the expansion or maintenance of general infrastructure. The first National Tourist Plan, completed in 1962, confined government plans mainly to expanding the national highway system, along with modernization of airports and areas surrounding existing tourist sites (Jiménez Martínez, 1990:94). Meanwhile the budget and prestige of the Department of Tourism remained quite modest. Its budget made up just .15 of one percent of total federal expenditures when it was created and remained close to that level for the next decade (Jiménez Martínez, 1990:93; Jud, 1974:26). With the exception of the 1968 Olympics and the 1970 World Cup soccer tournament, available funding would be confined to maintaining existing attractions and marketing abroad. Expansion of facilities was given little attention and almost no public resources.

By the mid-1960s the profile of Mexican tourism reflected these facts. Border tourism, which is not particularly lucrative, continued to make up the bulk of foreign arrivals to the country. A smaller percentage of foreigners made their way into the interior, primarily for purposes of business or cultural endeavors, but most concentrated on the capital. Coastal tourism flourished in Acapulco, but was otherwise negligible except for small, mainly Pacific coastal communities. Each of the segments of the industry developed largely in response to domestic and foreign demand and emerged without public planning or resources. To the extent that tourism grew, it did so with little tangible state aid.

Conclusion: Factors Facilitating Tourism Promotion

The above discussion demonstrates why by the mid-1960s Mexican officials saw a need to encourage exports, why an export push was viable, and why tourism was a somewhat obvious candidate. On the eve of the central bank study that called for rapidly developing Mexico's tourism industry, several significant developments had taken place. Domestically the economic model that had guided Mexican development for the past 40 years was coming under increased pressure. Although growth continued at steady levels, the flowering trade and budget deficits demanded greater attention. From 1960 to 1965 merchandise exports grew by three percent annually while imports grew by nine percent, and the overall current account deficit grew by 19 percent. While inflation was largely held in check, it still outpaced that of Mexico's major trading partners. Combined with a fixed exchange rate since 1954, this produced an increasingly overvalued peso (Reynolds, 1978:1007–1008). The model also resulted in increasingly unequal distribution of benefits. Underemployment flourished, and pressure for wage increases intensified. Social protest also grew, and perhaps more important from the standpoint of the government, it was spreading into the middle classes and into some of the traditional sectors of PRI support (Smith, 1991). These challenges were quite serious; but, as already argued, they were not yet at crisis proportions. Certainly if they were the bank would not have taken years to complete its final report. As a result targeting exports was a viable option for state managers.

Globally, international tourism was a booming economic activity, and was expected to continue to grow in the foreseeable future. High growth in First World countries generated more disposable income and leisure time, and simultaneously technological improvements meant the cost of traveling abroad was falling. Increasingly tourists were discovering the Third World, and as a result the industry was viewed more and more as a viable development tool. This was the case not only in developing countries, but also among First World elites concerned with development in the South. Yet, while tourism was becoming increasingly important to the Mexican economy, it had yet to be *actively* exploited by state officials. While the industry gradually became recognized,

state activity through this point had wavered under different presidential administrations and was passive at best. Instead growth in arrivals and receipts was
primarily a product of market forces. When addressed at all by state officials,
tourism exports were viewed mainly in instrumental terms, as a means to
finance industrialization. Few public resources directly targeted the industry
and instead the response to growing demand for tourism services was left
almost exclusively to private initiative. Tourism, then, was an obvious candidate for an export push in part because of its previous neglect.

Chapter 4

Initiating the Tourism Export Push:
The State Role

Mexican tourism has been marked by two major changes over the past 30 years: exports have flourished, and the *type* of tourism that foreigners now consume is different. No longer is the country primarily known for either border pleasures or "lo Mexicano"; instead it has become a major sun and sea destination. This chapter investigates the specific role played by the state in fostering change. The framework laid out in Chapter 2 indicates that features surrounding the state are likely to be crucial in shaping industrial transformation, both in terms of choosing the type of export scheme, and in concentrating the resources and political wherewithal needed to implement the project. This chapter demonstrates that rather than leaving development to market forces, tourism was in fact selected as an "export push" industry by state officials. They consciously altered incentives for private investors and took the lead in promoting its export. This leads to the additional questions of capacity and autonomy. To what degree did institutional arrangements, access to resources, and organization and opposition at the societal level affect the ability of state actors to carry out the export push?

The State and the Export Push

State efforts toward building up its tourism infrastructure began in earnest after 1968. It was then that the central bank, released ambitious plans calling for the vast expansion of tourism facilities in the country in order to increase export revenues and create jobs. The original central bank team was made up of 46 people, mainly bankers but also attorneys, architects, and urban planners. The initial study was begun in 1967 with $2 million in funding. The final product represented the ultimate in economic and social planning as well as reliance and a seemingly apolitical rationalism. According to Antonio Enríquez Savignac, who headed the team, "[a]s bankers, we approached this from a banker's point of view, taking everything measurable into account, feeding it into a computer, and leaving nothing to chance" (*The New York Times*, 1972:1). The project exhausted the central bank's computer capabilities and computer services were leased from a US provider. The studies called for building five new planned facilities, known as poles: Cancún in the eastern coast on the Yucatán peninsula; Ixtapa, not far from Acapulco in the state of Guerrero; Los Cabos and Loreto, both on the western Baja California peninsula; and the Bahías de Huatulco in the poor southern state of Oaxaca (Hiernaux Nicolás, 1989:111–112; INFRATUR/BANXICO; *The New York Times*, 1972). All would be beach resorts, all were completely planned from drafting tables located in offices in Mexico City, and all would be directed by state officials in what would be an export push.

The push was in some senses hardly surprising. As Chapter 3 demonstrates, the lack of adequate export revenues and regional development were chronic problems facing government officials by the mid-1960s. Dominant embedded orientations within the state also pointed toward a significant role for the state in any major development project. If anything this would increase during the 1970s under the Echeverría and López Portillo administrations, two presidents who oversaw a vast expansion of the public sector in the economy. Two other factors facilitated the ease of maintaining an export push, first Mexico's discovery of huge oil reserves in the mid-1970s which resulted in Mexico going from a net oil importer to one of the world's leading exporters, and second the growing availability of international capital from private banks during that same period. Both filled state coffers and would lead to ambitious state-led development programs throughout the economy in the 1970s and early 1980s.

State action in the tourism sector was indispensable in four particular areas: planning, provision of infrastructure, entrepreneurship and assumption of risk, and finance. In each area, state agencies took on a leading role, especially during the most uncertain periods. Inasmuch as state actors initiated a series of high profile, large-scale beach projects, it was they who ultimately defined what Mexico would become to the next generation of foreigners who would visit the country.

The State as Planner and Provider

The five poles or planned resorts held at least two distinct advantages for state officials. One was locational. Planning allowed for strategic placement of the resorts near targeted markets. This also served larger regional development goals in that all five poles were located in some of the least populated but poorest areas of Mexico. In addition, building resorts from the ground up was viewed as a means to overcome some of the more negative aspects of unregulated tourism development. Planning was seen as the key to maximizing the benefits while minimizing costs such as pollution, land speculation, hyperinflation, and rapid and unzoned growth that often led to the creation of large shanty towns near resorts. Planners and observers often compared the new poles to existing tourism areas, especially Acapulco, where unregulated tourism led to many of these problems (Ramírez Saiz, 1989).

Several institutional changes were undertaken during this time. A point agency for developing the poles, The National Trust Fund for Tourist Infrastructure (INFRATUR) was created in May 1969 (IMIT, 1980:19; Ramírez Blanco, 1983; Tancer, 1972). INFRATUR was located within — and administered by — the central bank. The agency was staffed primarily by bank officials, including several who undertook the original study of tourism. The first head of INFRATUR, for example, was Antonio Enríquez Savignac, who had earlier overseen the central bank study and came to be known as "the father of Cancún". Later he headed FONATUR, SECTUR, and finally the World Tourism Organization. Enríquez was something of a forerunner to the current generation of technocrats who have become dominant in contemporary Mexican ruling circles. He studied in Canada and received a degree at Harvard, then worked at the Inter-American Development Bank before moving to the *Banco de México* (Torruco Marqués, 1988).

In addition to responsibility for the poles, INFRATUR was to oversee the modernization of existing facilities and encourage private investment. The primary focus, however, was the poles. In Cancún, for example, INFRATUR was assigned the tasks of planning the site, building basic infrastructure, promoting private investment, developing and selling land, and maintaining overall oversight and project coordination with other government entities. In order to carry out these tasks, the agency was well funded and granted legal powers such as the ability to expropriate land (INFRATUR nd). The new bureaucracy did face some challenges, but timely intervention by Echeverría, who had overseen the Department of Tourism as part of his responsibilities as Interior Minister before becoming president, aided its success (Bosselman, 1978:40; Truett & Truett, 1982).

The fact that INFRATUR originated within the central bank is significant for several reasons. Most important, it indicates that tourism promotion came in response to balance of payments pressures and was conceived of mainly as an export project. In addition, many individuals who first staffed the agency came from the central bank, which itself was among the most orthodox institutions in the Mexican government. The bank was created in 1925 and while closely associated with private banking circles, has been relatively insulated from other state institutions and larger societal interests (Centeno, 1994; Maxfield, 1990, 1991:425, 430–437; Teichman, 1988). Historically, it has traditionally pursued a more conservative monetary policy than perhaps any other country in Latin America. Together with *Hacienda* (Treasury) and later the *Secretaría de Programación y Presupuesto* (Programming and Budget), the central bank was a body from which the orthodox technocratic revolution would later originate in the 1980s (Bailey, 1988; Centeno, 1994). Both the bank and INFRATUR staff were made up mainly of liberal economists, many of whom were educated abroad, especially in the United States. Moreover they shared a world view of rational problem solving that contributed to top down planning and implementation that would consistently mark state activities. It was the original vision of these central bank technocrats that led to the tourism push. In short, the embedded orientations within the agency combined export orientation with broader preferences for planning and a significant state role in shaping development.

The poles — again all planned to be quite large and located in coastal areas — were clearly aimed at foreign, middle class, mass tourists. The advantage was one of numbers. Beach tourism promised to draw throngs of foreigners deep into Mexico in what would hopefully be long, free spending vacations. This would also insert Mexico, however, into the center of the global tourism industry and one of the most competitive segments of it.

INFRATUR, along with FOGATUR, the trust fund created earlier to guarantee and subsidize tourism-related loans to the private sector, quickly began work to expand facilities in the new poles. The agencies were also greatly aided by access to cheap capital. NAFINSA, the national development bank, received a $21.5 million loan from the Inter-American Development Bank (IDB) in 1971 for the first stage of infrastructure development at Cancún (Acuña Jáuregui & de la Garza, 1989:124). The IDB loan for Cancún was the first direct financing made by the bank for a tourism project (Cowan, 1987:336). The total cost of the initial phase of development was projected at $47.1 million (IDB various publications; FONATUR nd). Later that year the World Bank announced approval of its first ever direct loan for tourism infrastructure in the form of a $22 million credit for the initial development of Ixtapa (Islas Guzmán, 1989:95; Tancer, 1975:31). Foreign financial support would become indispensable throughout the next decade, both for infrastructure and hotel construction. By 1981 the IDB had provided more than $300 million to Mexico for tourism development, including two additional large loans for Cancún. The first, made in 1976, financed much of the second stage of construction of the resort with $20 million. A second loan for $30 million in 1978 was aimed at expanding hotel facilities (IDB, 1986: 85; Jiménez Martínez, 1990:157).

State actors chose to concentrate first on Cancún and Ixtapa. INFRATUR devised and carried out master plans for each pole. Meanwhile FOGATUR, the lending agency that had been created in 1956, vastly increased its funding to private sector projects. Calculations based on data provided by Jiménez Martínez (1990:222) and FONATUR (1992) demonstrate that in 1973 the value of loans approved by FOGATUR for hotels shot up more than tenfold compared to the previous year and a year later the figure grew by an additional 300 percent. The money was earmarked mainly for hotel construction and went primarily to local private investors and the state-owned hotel chain Nacional Hotelera. This role of the state in channeling resources into the hospitality sector became especially important after 1970.

Pursuing large-scale tourist poles was justified by state actors on developmental grounds, including references to export earnings, job creation, and regional development initiatives. It should be noted that the large scale of the projects (plans for Cancún ultimately called for two million visitors a year, or nearly as many as visited all of Mexico in 1970) was consistent with dominant embedded orientations in most state ministries at the time. Also significant is that the specific emphasis on developmental goals varied in the case of each center. In Cancún, for example, INFRATUR argued the pole was needed to increase employment and regional development in order to alleviate regional migration to cities. In addition it would aid the balance of payments. However, the latter factor was clearly dominant (Hiernaux, 1999; INFRATUR, nd). The site itself, a nearly deserted barrier island located in one of the least populated areas of the country, was said to have been ultimately chosen after combining market analysis with computer-aided studies of weather patterns (*The New York Times*, 1972). Among the most important locational factors was the relatively untapped market for drawing US citizens to the Mexican Caribbean. When asked about the motivation behind the project Enríquez Savignac, then head of INFRATUR, replied, "Money. Tourists mean money." (*The New York Times*, 1972:1) Between 1961 and 1969 the number of Americans visiting the Caribbean had increased from 400,000 to 1.5 million, but the portion visiting Mexico was negligible. From the outset Cancún was aimed at that market and was thus clearly an export promotion project.

As Hiernaux (1999) points out, the location of Cancún was also based on geostrategic considerations and was closely linked to nation building. The area had few economic prospects after the end of the sisal boom and was adjacent to politically volatile Central America. Economic development was viewed as a way to alleviate social instability. In contrast, Ixtapa was originally planned as an attraction for both foreign and domestic tourists. The site was deliberately chosen to reduce pressure on increasingly overcrowded Acapulco. By 1972 the town was the only true world-class resort in Mexico and was drawing 1.5 million foreign tourists annually. Population growth had reached an alarming 20 percent each year (Bosselman, 1978:37; Jiménez Martínez, 1990:150; Ramírez Saiz, 1989:144–149).

Because the new resorts were built from the ground up in lightly populated areas, the tourism bureaucracy in effect became the governing power within the area. This was especially true during the initial phase of construction. In Cancún, for example, INFRATUR, which later became FONATUR, took on broad powers. The agencies expropriated land from the roughly 170 people living on the island and surrounding area, cleared the land, including some dredging of lagoons, and essentially erected a complete city (Bosselman, 1978). The first mayor of the resort town was previously director of FONATUR's community development office (author interview, 1992).

Similar plans were carried out in Ixtapa, where first INFRATUR and later FONATUR took on even broader powers. There a series of problems with local residents arose, especially in the already existing adjacent fishing village of Zihuatanejo. The small town, which did have some operating hotels before 1970, was primarily envisioned by tourism officials as a future worker support community for the larger resort of Ixtapa. Therefore, state officials planned to redesign the town in the expectation of rapid population growth. These initial plans, however, led to considerable dispute. FONATUR's initial plans, for example, called for completely rerouting streets in Zihuatanejo in order to modernize the town and prepare for increasing its population from 5,000 to 80,000.

FONATUR also initially expropriated about 9,000 acres in the area and created a trust to regulate land tenure. Conflicts over expropriation of communal lands (*ejidos*) and just compensation for the existing population produced conflict and led to significant construction delays before plans were put into effect. Expropriation of *ejidal* lands was particularly contentious (although nowhere near unique here) in that the *ejido* system was, until "reformed" in the 1980s, embedded in the Mexican Constitution and was to reserve these communal lands in perpetuity. As a result of the conflicts, construction in Zihuatanejo began in 1974, two years after that of Ixtapa (Cowan, 1987; Islas Guzmán, 1989:93–108; Long, 1991:207–208; Reynoso y Valle & de Regt, 1979:111–113). Again these episodes demonstrate the top-down, problem-solving ideology and practices utilized by state tourism offi-cials. This is consistent with Escobar's (1995) notion of imagining an underdevelopment that merely requires technical rather than political solutions. Although seemingly technical and rational in their efforts to promote tourism as a development tool, FONATUR officials were frequently seen as outsiders imposing their will on existing local communities through raw political power.

Aside from planning, one of the most important roles played by the state was providing tourism infrastructure. Federal government agencies, especially FONATUR, played central roles in virtually every aspect of building Cancún and Ixtapa. Here the state was instrumental in providing essential collective goods. At that time one of the biggest attractions of promoting tourism over other commodities or services was that it was viewed as a labor intensive and a "non-smokestack" or environmentally clean activity. Both, in retrospect, were mistaken conclusions. The start-up costs for such projects, for instance, were comparatively high (Daltabuit and Pi-Sunyer, 1990:9–13; Hiernaux, 1989). This was especially the case in mass tourist projects that centered upon sun, sand, and sea. They traditionally require a large-scale airport capable of accommodating wide-bodied jets, modern sewer and water facilities, electricity and other amenities that both tourists and private investors are accustomed to (and yet tend to be beyond the means of most residents of the country in question). In the Mexican case, the state provided this basic infrastructure.

In addition, it also took on projects that were viewed as necessary to attract foreign tourists, yet were expected to be unprofitable or profitable only over the long term. Foremost among these in the case of Cancún, for example, were building a golf course and central market while refurbishing nearby archeological sites (García de Fuentes, 1979; Hiernaux, 1989; INFRATUR, nd). All were considered indispensable for attracting both tourists and private hotel investment. In addition, the state planned and built Cancún City, a nearby worker city on the mainland. As Hiernaux (1999:129) states, this amounted to what would become a classic tourism model where the work and residential space of the employees would be segregated from the leisure space of the tourists.

It should be noted that environmental and social concerns received much less attention. Although state officials initially set aside one adjacent area to Cancún as an environmental reserve, it is clear that little attention was given to the impact that two million plus tourists in the area would have on the ecosystem. The same may be said for the social impact. The area surrounding Cancún has long been home to Mayan-speaking indigenous communities. Mass tourism certainly would have a profound impact on these communities as generations would later split between embracing modernity associated with tourism and larger capitalist development and retaining traditional ways (Pi-Sunyer and Thomas, 1997). Yet beyond focusing on dollar earnings, it is not clear that state officials made any attempts to anticipate the environmental and social impact associated with tourism development in this region.

Institutional Changes

Three further institutional changes enhanced the state's ability to intervene in the tourism sector during this period. First, the previously privately owned hotel chain Nacional Hotelera, which was on the verge of bankruptcy, was purchased by the government in January 1973 and restructured. At that time it operated just seven hotels with a total of 1,100 rooms. It also controlled several restaurants (IMIT, 1980; SPP, 1985:153). The chain would become an important state resource in the establishment of accommodation in the new tourism poles. Second, in passing the federal law to develop tourism in January 1974, the state funding mechanisms FOGATUR and INFRATUR were merged into a new agency, FONATUR, the National Fund for Tourism Development. FONATUR became the point agency for developing the planned resorts, as well as chief financier for other tourism projects. Third, later that year, the Department of Tourism was raised to *Secretaría* (a cabinet level ministry) in a move that gave it greater prestige and fiscal resources (de Mateo, 1987:186–187; IMIT, 1980: 22–23).

One logical question is why it was necessary to create two "point" agencies to guide tourism development. Technically, a division of labor existed between the agencies. As a cabinet-level ministry, SECTUR was the more overtly political of the two bureaucracies. It was quickly overshadowed by the more techno-cratic, better funded, and autonomous FONATUR, despite the fact that FONATUR is administratively under SECTUR. Some competition did emerge between the two entities, especially during the 1970s. FONATUR established a high profile as the central institution for developing the new tourism poles while SECTUR was bogged down in less prestigious tasks such as training of guides, compiling statistics, and marketing. FONATUR assumed broad powers, including those of land expropriation and resettlement of local inhabitants in the poles. It also directly carried out urban planning, construction of infrastructure, and marketing and promotion to private investors. Perhaps most important was its financial role. In fulfilling the earlier role of FOGATUR, the agency subsidized and guaranteed loans to the private sector for tourism-related projects (FONATUR, 1985).

Additional institutional aspects regarding the two agencies merit discussion. SECTUR was — and is — more politicized than its counterpart. It tends to be staffed on a more political basis, and turnover among the upper echelons of the ministry has often coincided with changing presidential administrations. Among recent Secretaries were Jesús Silva Herzog, who was Treasury Minister under Miguel de la Madrid and a serious precandidate for president in 1988 (losing out to Carlos Salinas); Carlos Hank González, the consummate behind-the-scenes PRI *político* who was once governor of the state of Mexico and later agriculture secretary; and Rosa Luz Alegría, who was widely alleged to be the mistress of president Jose López Portillo. Many ministers either previously or subsequently held elected political office or positions within the ruling PRI. The position of tourism secretary has served as a stepping stone to other offices, or in some cases constituted being put out to pasture. This is certainly not always the case, and is probably less so in Mexico than elsewhere in the region, but when it does happen it is consistent with Schlüter's position that in much of Latin America "... [w]ith very few exceptions, the highest executives in the tourism area are appointed for political reasons ... They are expected to provide political responses and not technical ones" (1994: 255).

In contrast FONATUR has been somewhat more meritocratic. Staffers historically have come from the central bank, other finance-related ministries, or the private sector. According to one long-time official with experience at both FONATUR and SECTUR, the former was staffed primarily with personnel drawn from the central bank or private sector while the latter drew more from political circles (author interview, 1992). This is not to suggest that these divisions are set in stone or that FONATUR and SECTUR are completely separate. Enríquez Savignac, who was the first head of INFRATUR and later FONATUR, was SECTUR minister from 1982 to 1988. Mario Ramón Beteta, Director General of FONATUR in 1992, was also active in PRI party politics, having once served as a governor, in the Treasury Ministry, and the state-owned oil monopoly (author interview, 1992). Beteta is also son of Ramón Beteta, who was a powerful political figure and served as Mexican Finance Minister under Miguel Alemán from 1946 to 1952. The point remains, however, that FONATUR is more careerist and continues to be linked to the agency's central bank origins. As a result, FONATUR also has also been seen as an agency of considerable prestige.

Second, FONATUR gradually became something of an independent entity in itself. In the past 25 years, the agency was moved from being under the administrative control of the central bank, to NAFINSA, and finally to SECTUR. Despite these changes, however, the agency has continued to enjoy considerable independence and autonomy. Most important, it has become largely self sufficient, or at least has had an independent source of resources for much of its history through its access to significant funding from the World Bank and IDB. Between 1971 and 1986, the two institutions provided loans totaling $471.5 million to Mexico for tourism development. All of that flowed through either NAFINSA, FONATUR, or both (Schédler, 1988). In addition, because it acts as a bank, the agency has gained significant independence from the larger bureaucracies of which it has technically been a part. Finally, FONATUR has made use of its power to buy land, make improvements upon it, and then sell it for profit. All of these generated some degree of financial autonomy.

The net result was that regardless of changes in the bureaucratic chain of command, INFRATUR and later FONATUR enjoyed considerable independence from internal political forces, especially at early stages in the export push. It also commanded the resources and power necessary to implement its particular vision of tourism. That vision was one aimed squarely at the export market, relying on planning and scale, with the state itself leading the way.

The fact that tourism was in its infancy also benefited the agency in that it did not face powerful societal opposition to its early plans. The only clear demonstrated opposition the state faced was that of small local groups surrounding big tourism projects who lacked significant power. Most was confined to groups of small landholders and poor *ejidatarios*. Because the groups were small and state actors enjoyed strong legal powers, the conflicts almost invariably were settled on terms imposed by the state (Cowan, 1987; Long, 1994; Reynoso y Valle & de Regt, 1979; Shacochis, 1989). The state was able to carry out the pole concept because of a striking absence of action from sectorally-organized private interests. This can be traced mainly to the previously dominant type of tourism prevalent in Mexico during the 1960s and 1970s. While growing rapidly by the 1960s, tourism had been confined mainly to the major cities or border areas. The sun and sea segment of the tourism market was underdeveloped. To the extent that hotel owners and other tourism related service providers were organized, they saw little threatening competition from the poles and some believed they would benefit from the expansion.

As a result, in this case of export push state actors were not confronted by strong, industry-based interests that either had to be overcome or accommodated. Instead, as the discussion below demonstrates, the biggest problem confronting state actors was enticing private interests into the new tourism zones. Only with time, and growth of the tourism industry, did private actors become much more involved in the sector and more organized. Therefore, state actors in key agencies developed the capacity, prestige and necessary resources, further, they faced few societal constraints in putting their vision of tourism promotion into action.

The Fruits of State Action

The rewards of the planned resorts began to arrive as early as 1974, when the first foreign tourists arrived in Cancún. As Table 4.1 demonstrates, more than

Table 4.1: Arrivals to Cancún (1975–1991).

Year	Arrivals[a] (thousands)	Variation (%)	Nationals (%)	Foreigners (%)
1975	99.5	–	72.5	27.5
1976	180.5	81.4	62.9	37.1
1977	265.2	46.9	56.0	44.0
1978	309.8	16.8	51.7	48.3
1979	395.9	27.8	49.5	50.5
1980	460.0	16.2	47.5	52.5
1981	540.8	17.6	48.8	51.2
1982	643.8	19.0	47.8	52.2
1983	754.7	17.2	32.4	67.6
1984	713.9	–5.4	30.0	70.0
1985	729.9	2.2	31.1	68.9
1986	869.3	19.1	26.2	73.8
1987	960.6	10.5	20.1	79.9
1988	838.2	–12.7	21.6	78.4
1989	1153.6	37.6	25.7	74.3
1990	1575.7	36.6	25.1	74.9
1991 p/	1912.1	21.4	25.1	74.9

[a]Includes tourists staying in hotels.
Source: Calculated using data from FONATUR, reported in SECTUR (1992).

27,000 visited the island in 1975, which marked the completion of the first of three phases of development (Bosselman, 1978:44–45; Hiernaux, 1989:113). From that point on Cancún started down the path of becoming a world-class resort destination. The number of hotels increased from nine to 42 between 1975 and 1979, while room totals doubled (Acuña Jáuregui and de la Garza, 1989:124). Between 1975 and 1984 international arrivals to the island grew by an average annual rate of 38 percent and by the mid-1980s the destination accounted for more than one-tenth of all foreign tourists to Mexico (Hiernaux, 1989:113; SECTUR, 1985:42). In 1986 Cancún drew more foreign tourists than Acapulco for the first time, and by 1989 it passed the Federal District surrounding Mexico City to become the single most popular Mexican destination for international tourists (SECTUR, 1991:43, 161, 167, 172, 187).

Similar if less spectacular results were subsequently recorded at the other four poles. Ixtapa-Zihuatanejo began attracting foreign tourists by the mid-1970s. On

Table 4.2: The Poles and International Tourism Exports.

Year	Int'l Arrivals to Poles[a] (thousands)	Pole/Total Int'l Arrivals (%)
1975	39.8	1.2
1976	79.9	2.6
1977	133.5	4.1
1978	179.2	4.8
1979	246.0	6.0
1980	332.6	8.0
1981	382.0	9.5
1982	480.3	12.8
1983	762.6	16.1
1984	783.2	16.8
1985	751.6	17.9
1986	903.0	19.5
1987	1088.5	20.1
1988	988.0	17.4
1989	1202.7	19.4
1990	1585.6	24.8
1991	1873.9	29.4

[a]Poles include Cancún, Ixtapa-Zihuatanejo, Los Cabos, Loreto, and Bahías de Huatulco.
Note: International Arrivals to Poles includes only those who stay in hotels.
Source: Calculated using data from SECTUR, *Estadísticas Básicas de la Actividad Turística*, (various issues).

the Baja California Peninsula, Los Cabos and Loreto came on line in 1980 and 1981, respectively, and finally Huatulco began attracting foreigners in 1987. Table 4.2 summarizes the growth and increasing importance of the five developments over time. The lone year in which the share held by the five poles failed to grow, 1988, was due primarily to hurricane Gilbert, which struck Cancún that year. It shows that by the 1990s — 15 years after they began receiving tourists — the centrally planned resorts together accounted for about one quarter of all international tourist arrivals to Mexico. In fact, available data tend to underestimate the importance of the planned resorts. The totals for the poles only include hotel guests, thereby ignoring those who stay in condominiums, second homes, time shares, or private homes with friends or family. In other words, in the only year in which such data are provided, SECTUR estimated that of the 6.37 million international tourists who came to Mexico in 1991, 4.75 million stayed in hotels. Of that total, 1.87 million or 39.4 percent of the hotel stays were in the resorts planned and built by FONATUR (SECTUR, 1991:161). The proportions are also obviously higher if one considers segments of the market such as sun, sand, and sea tourists. In 1991, for example, the poles accounted for 62.4 percent of arrivals staying in hotels in Mexico's largest beach resorts.

In addition to creating the new development poles, SECTUR and FONATUR were active in a series of other projects. SECTUR stepped up marketing activities abroad by opening several new offices. Both agencies also intensified efforts toward drawing tourists to traditional resorts as well as to new sites. The experience of Puerto Vallarta, on the Pacific coast, is indicative of these efforts. There infrastructure projects included building a paved highway, international airport, marina, and deep water pier. These helped make the town an international destination. In 1973 a special *fidiecomiso* (trusteeship) was created to regulate land (Chant, 1992; Evans, 1979). The federal government also created several trusts that provided matching promotional funds for individual tourism locales in the country (author interview, 1992). Further, SECTUR undertook a series of actions to facilitate the education and training of personnel working in the sector. SECTUR, in particular, was active in supporting and creating tourism schools (Jiménez Martínez, 1990:176–177). Perhaps most notable, during the late 1970s and early 1980s the agencies emphasized planning beyond the five development poles to the nation at large. This planning took a number of different forms, from increased efforts to coordinate agencies and state-owned enterprises, to zoning and cataloguing all attractions in the country (Jiménez Martínez, 1990:169–174; Ortiz de la Peña Rodríguez, 1981).

As a result of all these activities, the state assumed a new role as the primary initiator and overseer of tourism development in Mexico. While the promotional aspects of state action were hardly revolutionary, planning tourism resorts explicitly for export purposes was significant and quite new. In pursuing the poles, the bureaucracies were instrumental in changing the image of Mexico. The planned resorts did help avert some of the problems associated with market-driven tourism, but planning itself opened the door to criticisms centering upon sterility and an artificial aura surrounding the resorts (Daltabuit & Pi-Sunyer, 1990). Perhaps more significant, the poles had the effect of bringing the sun and sea segment to the forefront of Mexican tourism. All five of the planned poles were coastal attractions, and were marketed as such. By the early 1980s the "traditional" Mexican tourism centers of Mexico City, Guadalajara, and Monterrey were drawing less than 30 percent of international arrivals to the country (Molinero Molinero, 1982:36). Ironically, by the late 1980s state officials were actively attempting to diversify attractions, promoting, for example, Colonial Cities and the Mayan Route. As one tourism official said, "Many people think of Mexico as only having beaches" (author interview, 1992). Instead, for many foreigners, the Mexico they came to know and identify with became Cancún or another beach destination.

In examining industrial transformation from the standpoint of the role of the state, it is not enough to show that state agencies were active, had resources and room to maneuver. Also necessary is demonstrating that state action somehow altered outcomes. In other words, what is the counterfactual involving tourism development in Mexico had the state not intervened? The alternative is uncertain, but clearly resorts such as Cancún would not exist. The second way in which the state "mattered" is that it played a crucial role in intervening in the market to shape incentives for private actors, and also performed tasks that private-sector actors were unwilling to take on.

The State as Risk Taker and Financier

Aside from shaping the general trajectory of Mexican tourism, the state also provided two other important functions: it assumed most of the initial risks associated with getting the poles off the ground and it financed the vast expansion of hotels in the country over the next two decades. In this sense, this action was consistent with Bennett and Sharpe's (1982) notion of the state as banker and entrepreneur. In other words, the state's activities were not completely unique to tourism. During this period, the state provided much of the gross fixed capital formation and took on a growing entrepreneurial role, factors Bennett and Sharpe link to specific developments in peripheral capitalism. With respect to tourism, the state took on these tasks in the more specific context of export promotion. Aside from paying the costs associated with the basic infrastructure for the planned resorts, the state also took on the entrepreneurial risks inherent in developing what Erdman Gormsen (1989:83) refers to as a tourist "pioneer zone".

The central problem facing state actors was that despite provision of infrastructure, private actors were for the most part unwilling to risk investing in a new tourism zone. FONATUR responded by entering the hospitality business through acquisition and expansion of the hotel chain Nacional Hotelera in 1973. One SECTUR official, in referring to the difficulty of drawing investors into the new poles, put it this way: "... not one hotel chain wanted to operate a hotel in a place where there was nothing" (author interview, 1992). The chain was put directly under the administrative control of SECTUR when it became a state entity in 1972, although it was primarily operated through FONATUR (Molinero Molinero, 1982). FONATUR built new hotels in the poles and the chain operated them. Primarily through its investment, the chain expanded from seven to 28 hotels, operating under the commercial name of Presidente, by the time it was re-privatized in 1985 (Jiménez Martínez, 1990: 187; SPP, 1985:153). The state also inherited the chain's restaurant and nightclub holdings in 1973. In the interim, the state used the enterprise to overcome collective action problems in the planned resorts. Presidente hotels were among the first to open in Cancún, Ixtapa, and Loreto.

FONATUR pursued a similar strategy in choosing to develop all-inclusive resorts in the new poles. The agency reached a franchising agreement with Club Méditerranée, and then built, owned, and operated Club Med resorts in Cancún, Ixtapa, and later Huatulco. Again, the all-inclusive resort-hotels were among the first establishments to open in the poles, and were targeted to attract tourists into the new zones shortly after the airport became operational. The all-inclusive hotel-resorts served the dual purposes of making the poles more attractive to investors and bringing tourists into an area that still lacked many support services and other tourism activities. Usually restaurants, night clubs, and other attractions were not yet opened. As one state official put it, "Club Med was necessary for us to create tourist traffic" (author interview, 1992). An all-inclusive resort was figured into the original plans for Cancún (and perhaps other poles) (García de Fuentes, 1979:89–90). Further, FONATUR often built hotels that were sold during various phases of construction and operated by the private sector.

In addition to its entrepreneurial role in building and operating hotels, the state also vastly expanded its efforts toward financing hotel construction. FOGATUR and later FONATUR altered market incentives for potential investors by establishing an extensive program for offering preferential credit for lodging to private investors. The program was primarily for new hotel construction, although it also included credits for renovation and expansion of existing facilities. Additionally, a small amount of financing for restaurants and bars was provided. FONATUR was also able to target where the hotels went, and agency loans were crucial in channeling investment toward the poles. Again FONATUR itself was aided in this process both by gaining access to international monies. Between 1978 and 1986, the World Bank and IDB made available almost $300 million to Mexico for hotel construction, with FONATUR administering the loans (Schédler, 1988:174). Based on FONATUR data reported in SECTUR (1986), it appears that during the first ten years of operation, FONATUR offered financing to the private sector totaling $1.5 billion.

FONATUR guaranteed hotel loans through national financial institutions and also subsidized the cost of money by offering preferential interest rates and amortization (FONATUR nd; Jiménez Martínez, 1990:177–178; Ortiz de la Peña Rodríguez, 1981:21–24). The maximum amount of the loans was capped. Although interest rates also varied by hotel category, typical terms were for repayment to be made over 15 years with a three-year grace period (FONATUR nd:5; Ortiz de la Peña Rodríguez, 1981:21). Calculations from agency records reported in SECTUR (1986) show that between 1974 and 1986, the first twelve years of stepped up lending, average interest rates of state loans were 19.6 percent lower than market rates. The subsidies also increased during that period. From 1974 to 1978, rates were subsidized by an average of 15.4 percent, while between 1982 and 1986 the average subsidy was 24.8 percent.

After 1974 the FONATUR loans provided a major impetus to hotel construction in the country. As Table 4.3 shows, state credit offerings increased steadily after 1972. Between 1974 and 1992 FONATUR financed the construction of more than 100,000 new hotel rooms. In other words, state money built nearly a third of the total number of hotels that exist in the country today (Chant, 1992:89; FONATUR, Dirreción de Crédito nd; SECTUR, 1991). In some years, FONATUR accounted for between 50 and 75 percent of the total public and private investment in tourism (Truett and Truett, 1981:14–15). The agency also provided loans to remodel existing hotels and construction of condominiums, houses, and timeshare installations. FONATUR concentrated on hotels, however. The number of condominium rooms financed by the agency totalled just over 4,000, while timeshare unit totals were 2,500 and houses just under 600 (FONATUR, 1992).

Over time, FONATUR lending has been augmented by financing from other sources, most recently by BANCOMEXT, the state export-import bank (author interviews, 1992). In recent years, the bank has been aggressively involved in tourism development, including that of Megaprojects, the new "super-resort" areas currently under construction. Earlier, in 1982, the state had created a special, short-lived National Tourism Bank (BANTUR). Operating under SECTUR, it was to finance both supply and demand for tourism (SECTUR, 1983:38, 59–60; Torruco Marqués, 1989:169–173). It is not clear if BANTUR was meant to somehow compete with FONATUR. The bank was very much a pet project of president Jose López Portillo and his Secretary of Tourism and widely alleged lover Rosa Luz Alegría. According to one FONATUR official, the bank only made one loan before being disincorporated by López Portillo's successor, Miguel de la Madrid (author interview, 1992).

Despite a sharp but brief drop off in state lending coinciding with the debt crisis in 1982 and 1983, FONATUR financing was indispensable over the past

Table 4.3: Hotel Financing by FONATUR (Millions of Current dollars).

Year	Amount	Year	Amount
1957	0.10	1975	66.14
1958	0.92	1976	48.27
1959	0.50	1977	47.41
1960	0.50	1978	104.21
1961	2.77	1979	112.74
1962	7.33	1980	225.05
1963	2.10	1981	367.14
1964	6.66	1982	78.49
1965	1.83	1983	44.49
1966	1.85	1984	162.14
1967	3.67	1985	113.18
1968	5.85	1986	76.88
1969	3.32	1987	57.75
1970	2.27	1988	7.50
1971	3.07	1989	81.86
1972	2.31	1990	42.36
1973	25.11	1991	86.26
1974	73.18	1992[a]	42.57

[a]Through June 30, 1992.
Source: Author calculations based on: 1957-1973, FONATUR data reported in Jiménez Martínez (1990:222); For 1974-1992, FONATUR, Dirección de Crédito; IMF, *International Financial Statistics Yearbook*, various issues.

20 years for expanding and upgrading hotel facilities in the country. By the beginning of the 1990s Mexico ranked eighth in the world in hotel room offerings (SECTUR, 1991). Breaking the aggregate lending totals down over the entire period reveals two additional points. First, state financing was crucial for hotel development in the poles. Data from FONATUR (Dirección de Crédito, through June, 1992) show that between 1974 and 1992 more than one-third of all credits went to the states of Quintana Roo and Guerrero, the respective sites of Cancún and Ixtapa. These totals were even higher during the first stages of building the planned resorts. The loans helped Cancún to rank second nationally in the total number of hotel rooms by 1990. The resort totalled 17,470 rooms in 1990, about 700 fewer than the Federal District. By way of comparison, the country of Jamaica, the leading destination in the Caribbean, had 16,000 rooms

in 1990, while Hawaii totalled 72,000 (SECTUR, 1991). In recent years a higher proportion of FONATUR financing has gone to the newer poles in Baja California and Oaxaca.

Second, breaking down the financing of projects by the size and quality of hotels demonstrates that state lending practices increasingly favored the export market through concentrating on upscale, international-quality offerings. In Mexico, as in many nations, hotels are graded by stars, ranging progressively in quality from one to five stars, and finally "Gran Turismo". Most international quality hotels fall within the categories of Gran Turismo or five stars, although four-star and some three-star hotels are often linked to international chains. Several are classified as "special category, economic class and without classification" (SECTUR, 1992:16). As of that year Mexico had 345,159 rooms: Gran Turismo and Special Class five percent, Five Star ten percent, Four Star 13 percent, Three Star 14 percent, Two Star 13 percent, One Star ten percent, "Economical" Class 13 percent, and Unclassified 19 percent (SECTUR, 1991:315).

With respect to the categories, FONATUR's operations increasingly subsidized hotels that would cater to foreigners. Although data are incomplete, evidence from 1984 to 1991 shows that among the top three categories, the number of hotel rooms grew by 67 percent. During the same period, the total number of rooms in Mexico grew by just 31.5 percent (SECTUR various issues). According to one SECTUR official, FONATUR financed roughly 85 percent of hotel construction during that period, suggesting that luxury hotels drew the bulk of state funds (author interview, 1992). As the next chapter demonstrates, most of these hotels were linked with international chains and/or large Mexican business conglomerates.

This pattern of focusing on larger, more capital intensive projects is consistent with a more recent state initiative, a series of new developments called Megaprojects. Early during the presidential administration of Carlos Salinas (1988–1994), FONATUR and SECTUR jointly announced plans for the Megaprojects, which are in effect self-contained mini-resorts that involve lodging, some transportation, and recreational services. The projects, which number between nine and 20, follow FONATUR's preference for planned resorts, but are aimed at diversification in an increasingly segmented market. The Megaprojects were initiated in 1989 in the context of a National Modernization Project for Tourism (SECTUR, 1989, 1994). Originally twelve projects were announced. The total number has fluctuated since then.

As a result, many of the projects are aimed at higher-end tourists, while others attempt to capture the growing market for camping, sport fishing, or short-term and more modest border travel. The bulk of the Megaprojects, however, are directed at those tourists who are especially affluent. Many are gated enclaves that have little connection to the surrounding area. Several involve building marinas that accommodate large yachts. Puerto Cancún, for instance, a nautical center currently under construction near the existing planned resort, calls for building a wharf with 500 slips for yachts, a golf course, and luxury hotels and condominiums.The plan calls for a series of navigable canals, modeled somewhat like a residential Venice (SECTUR, 1992:60). According to one official, hotels that charged less than $300 per night would not be allowed in Puerto Cancún (author interview, 1992). Another FONATUR spokesperson Carlos Zugasti Islas also indicated that Megaprojects are aimed at attracting "higher quality tourists" (*Mexico Journal*, 1989:20) suggesting this path allows Mexico to compete with the European market.

Perhaps more important than the size and scale of the Megaprojects is the manner in which they are to be built and financed. Again FONATUR plays a central role by identifying locations and taking the lead in planning the physical site. In addition the agency is offering a menu of options available to investors for developing the mini-resorts. They range from initial privatization for development, to state development of the infrastructure, and then selling to the private sector, to FONATUR acting as a full partner in all phases of the project. The more important point, however, is that only large-scale private investors have access to the projects due to the manner in which concessions are made. For example, original plans for Puerto Cancún slated overall costs at $1.5 billion. The method of concessions has been to offer either partnership or complete development to the private sector, but the state will only grant concessions for development to a single group. Meanwhile the state maintains the leading role in planning, infrastructure, financing, and entrepreneurship.

Conclusion: The State and Industrial Transformation

This chapter demonstrates that state actors stood out as active participants throughout the tourism push. One simply cannot argue that tourism dynamism took place due to market forces alone. Aside from documenting just what state actors did, there are also important theoretical reasons for focusing on the roles played by public institutions. Chapter 2 argues, that first, export pushes are likely to result when external stimuli are favorable *and* when embedded orientations held by public sector actors favor statist solutions to development problems. It is clear that with respect to tourism this was the case. Although state managers may have been more orthodox than some of their counterparts, they made it clear that tourism promotion would be planned and directed from the public sphere.

A second theoretical claim is that export pushes that violate market ortho-doxy may be successful if state attributes favor formulation and implementation of plans and if societal constraints can be overcome. State actions did violate market signals in this case. Public agencies and firms helped overcome collec-tive action problems by stepping in at critical junctures when private actors were unwilling to act. The state assumed significant risks in order to attract private investment into the poles and also offered attractive financing to inves-tors. Internally a series of reforms created powerful, insulated, and cohesive bureaucracies that gained access to the necessary resources and policy instru-ments to successfully implement plans. In addition, the requirements for instrumental autonomy were met. Societal opposition to state plans was rela-tively small and confined to weaker, geographically dispersed groups. Certainly tourism promotion plans were not confronted by leading capitalist groups in the country. In part this was because tourism was not well developed in Mexico, although also beneficial was foreign financing that was earmarked for the industry. As a result, at least during the 1970s and early 1980s, state actors enjoyed considerable freedom in planning and implementing the tourism export push.

As a result, Mexican tourism changed markedly over the past 30 years. Growth has been rapid, but beyond size alone, the most notable change, for better or for worse, is that most tourism now revolves around beaches. This is a direct result of state planning along with the other activities documented above. Concentrating on the poles relieved pressure on the traditional destinations, while creating employment and funneling export earnings into several of the poorest areas of the country.

While state action was critical in these areas, it is not argued that in this case the state was a lifeboat of efficiency in an ocean of red tape. Rumors and stories of mistakes, corruption, and other inefficiencies surrounding tourism develop-ment are not uncommon. Various politicians, friends of politicians, and those with political ties have certainly benefited from the growth of the industry. It is interesting to note, however, that if anything the stories of such activity have grown with the wave of privatization in recent years. It is unlikely that state tourism agencies were "above politics", and in fact political ties may have enhanced their effectiveness. Nor is it to suggest that state planning completely averted problems. Today SECTUR and FONATUR officials openly admit they overbuilt Cancún, for instance. Indeed many environmental and social problems may be becoming particularly serious.

Equally necessary is to consider which factors are ignored by policymakers when they set goals. At the top of the list were environmental and social concerns. Today the carrying capacity of Cancún and the adjacent Costa Maya to the south of the resort is severely strained by heavy tourist traffic. Sewage problems, urban congestion, water pollution, and threats to native plants and animals have become serious problems around many of the poles (especially Cancún) and ancient ruins. Land use and environmental impact studies have been either completely ignored or at best secondary considerations when implementing tourism promotion plans (Pi-Sunyer & Thomas, 1997:197). This is particularly ironic given that both sand-and-sea tourism and tourism that focuses upon historical sites or nature rely upon fragile and non-renewable resources. Yet among state planners nature was viewed as something to be conquered and historical sites to be exploited in order to produce export revenues. This reflects a larger economistic and technocratic ideology that has dominated the overall conceptualization of tourism export promotion by state elites for the past 30 years.

Despite these problems, the institutional requirements for what state officials themselves defined as a "successful" export push were in fact met, and in their own terms tourism is extremely successful in Mexico from a developmental standpoint. Certainly the central goals of policymakers, increasing export earnings and providing regional employment, have been largely achieved. The next two chapters, however, turn to distributional aspects of tourism development and show that there the record is more mixed.

Chapter 5

Patterns of Hotel Development

While the previous chapter summarized the growth aspect of industrial transformation, this and the next one offer detailed examinations of tourism sub-industries in order to highlight the second component of industrial transformation. Tourism has been successful in terms of export growth, but what of the distributional consequences of that growth? Equally important, what determines ownership and control patterns within export industries? This discussion traces the evolution of the global and Mexican hotel sectors, demonstrating that industry characteristics and resulting international structures tend to be reproduced on a local scale. The outcome in hotels is a complex web of ownership and contractual agreements between global hotel chains and internationalized Mexican capital.

Hotels are of central importance in the analysis of the political economy of tourism. Because official statistics omit air travel from aggregate international tourism totals, the bulk of those figures actually reflect lodging expenditures. This suggests that among the primary beneficiaries from tourism are owners and operators of motels, hotels, guest houses, and other lodging facilities. In addition, their workers make up a significant portion of most aggregate figures on tourism employment. For instance, 1988 SECTUR figures show about 14 percent of direct employment in tourism is made up of hotel workers. This represents the third highest total after "miscellaneous" and restaurants and bars (Hiernaux & Rodríguez Woog, 1991). Hotels merit particular attention because they lie at the heart of the accommodation sector. As already demonstrated, hotels have grown rapidly in Mexico over the past two and a half decades, as the country added more than 100,000 rooms to its stock.

The argument pursued here is basically that two particular aspects shape the formation and use of industry assets in the hospitality sector. First, consumer uncertainty creates a heightened need for trust or reliability, a factor particularly common to services. Product reliability, of course, is also important for goods, but there is a premium placed on this factor in services because consumer services produce an experience. Experiences cannot be exchanged as one might exchange stale bread or a television set that does not work. One can be compensated, but such compensation must come after the fact. In short, consumption of a "poor experience" cannot be corrected; it may only be made up for. The essence of the product is, in fact, the quality of that experience. This nature of the hospitality product, in addition to other requirements such as capital and technology, lead to the creation of firm-specific assets in the reputation for reliability. Second, hotel chains have gained the ability to separate these assets from actual property ownership. Each of these factors becomes crucial for understanding the unique nature of the international hotel sector, and in turn, its industrial transformation in Mexico. These and related issues need to be discussed and understood.

The Hospitality "Product" and the International Hotel Sector

The hotel business constitutes a unique economic activity in that it is really two businesses: providing hospitality services and real estate. The two were once combined, but became separable with the appearance of chains. Hotels, like many other global tourism sectors, began to form a more clear organizational structure after World War II. Prior to the war most hotels and motels were independent operations. The first US chain may be traced to entrepreneur Ellsworth Statler, who opened his first hotel in 1908 in Buffalo before expanding into Boston, Cleveland, and other cities (EIU, 1987:44). But chains were rare. Instead, owners were operators, and they mainly catered to business tourists. In the United States, less than five percent of hotels were associated with chains in 1948, while the worldwide portion was slightly higher (EIU, 1988:27, 30). After the war, however, the sector was marked by an increasing amount of association in chains, and by internationalization. Conrad Hilton began buying small hotels in the interwar period and eventually bought the Statler chain. The first Sheraton hotels, a run-down set of New England properties, emerged in the 1930s but did not take off until after World War II (EIU, 1987). The first Holiday Inn was opened in Memphis in 1952. By the early 1960s, this chain was opening new hotels at the rate of one per week and passed Sheraton as the largest hotelier in the world. Because they centered on business tourism, chains established most of their hotels in metropolitan areas, usually in city centers. Hilton, Sheraton, and Inter-Continental were the first well-known chains, while Holiday Inn began as a motor inn and only later gradually expanded into the urban hotel business (EIU, 1988). Only later, as pleasure tourism became more popular, did it move aggressively into resorts and abroad.

Industrial Organization theory suggests that when contemplating entry into a foreign market, transnational companies (TNCs) face initial disadvantages *vis-à-vis* local enterprises, including a lack of knowledge of the market and less flexibility. Therefore, they tend to enter those markets only when the disadvantages are overcome by certain competitive advantages (Dunning, 1970; Hymer, 1976). In hotels, capital requirements, along with the necessary technological, managerial, marketing, and organizational skills, are not considered to be highly complex, nor are they inseparable from ownership, as in many manufacturing industries (UNCTC, 1982). As a result, TNCs lack clear oligopolistic advantages, at least in the traditional sense. This would suggest that local firms would have little trouble gaining access to local hospitality markets. Some argue, however, that international chains may have expanded abroad not because of oligopolistic reaction, but rather simply to follow their customers, mainly business tourists who increasingly took their business abroad (Ascher, 1985; UNCTC, 1990:98; United Nations Transnational Corporations and Management Division, 1993). Finally, some chains were owned by airlines and appear to follow route expansion. For example, Inter-Continental hotels were created by Pan Am in 1945, opening their first hotel in Brazil (Ascher, 1985:24; Dunning & McQueen, 1982:74). TWA owned Hilton International and American Airlines owned Flagship and Americana Hotels. Later Hilton International and Westin (originally Western) were linked with United through their then parent Allegis (Bull, 1991:186–187; UNCTC, 1988:402). Historically chains formed for many reasons, and similarly expanded abroad out of different strategies. Examination of firm practices, however, demonstrates that hotel chains tend to behave as oligopolists (Bull, 1991; EIU, 1987:46). What they hold in common is a particular asset that they brought with them in expanding into new markets: trust.

Uncertainty, Strategic Assets and the Growth of Chains

Small but cumulative advantages may give hotel TNCs a somewhat favored position over local firms (UNCTC, 1982:47, 1990:90). Computers, and computer reservation systems (CRS) certainly amount to a more traditional source of monopolistic advantage for TNCs. Hotels, like airlines, were among the first to utilize computer technology for reservations, pricing, and back room uses. The primary source of oligopolistic structures and behavior, however, emerges from the reputation for trust or reliability, which is embodied within the very nature of the product itself (Bennett & Radburn, 1991; Lanfant, 1980; Lattin, 1990:219–223). As is the case with many (but not all) services, a stay in a hotel room is an "experience good". Unlike most goods, it cannot be thoroughly inspected before being consumed (Dunning & McQueen, 1982:83; Witt, Brooke & Buckley, 1991:61). Experience goods contrast with "search goods", which can be inspected or examined more closely before purchase. Because they cannot be fully inspected, potential hotel customers seek to find ways to contain this extra risk. One such way is to rely on firm reputation. In other words, trust may be embodied in a brand name, and that name makes a particular difference in the case of hotels. As Dunning and McQueen argue, "this is a powerful competitive advantage" for firms (1982:83). Reliability initially created incentives for the formation of chains, and also encouraged chains to expand abroad. This firm-specific asset becomes especially powerful where customers are in an unfamiliar environment such as a foreign country. In a sense hotels provide more than trust in this sense. Many who travel abroad as mass tourists have little knowledge of the destination country and along with uncertainty they frequently possess some degree of fear. Increasingly hotel chains provide an environment of familiarity by providing a range of services. These include a tourist class room, familiar language, spatial arrangements, and provision of consumer goods. Hotels become, as much as a place to sleep, a familiar and comfortable refuge for foreign tourists. While most may claim to be adventuresome, when it comes to accommodations they favor a name (and frequently environment) they know.

Trust, embodied in a brand name, becomes carefully cultivated by chains to the point where it becomes integrated with the product itself. Part of what is paid for in this experience good is the experience of peace of mind. Reliability gives well-known chains in tourism sending markets intangible yet clear advantages over local competitors in the receiving market. Uncertainty and reputation help to explain the growth of TNC chains over the past 25–30 years. Table 5.1 summarizes the top 20 hotel TNCs as of 1999. It shows that eight or nine of the top ten largest firms are US based and all are based in AICs (Bass is based in England but its hotel holdings are headquartered in Atlanta). In fact depending on how development is defined, either 49 or all 50 of the top 50 TNCs are located only in these countries. One TNC, CDL Hotels (ranked 45th), is based in Singapore (*Hotels*, 1999a).

The most significant work on the structure of hotel TNCs has been done by Dunning and McQueen (1982), who also were principal authors of UNCTC 1982. Their research shows that the accommodation sector is becoming increasingly concentrated as chains expand worldwide. In addition, chains themselves are often associated with or owned by other providers of tourism services. Within their sample of 81 hotel TNCs, 16 were associated with airlines, and six more were affiliated with tour operators and travel agencies

Table 5.1: World's Largest Hotel Chains (1998).[a]

Rank	Firm	Country	Rooms	Hotels
1	Cendant Corp.	USA	528,896	5,978
2	Bass Hotels	USA	461,434	2,738
3	Marriott Int'l	USA	328,300	1,686
4	Choice Hotels	USA	305,171	3,670
5	Best Western	USA	301,899	3,814
6	Accor	France	291,770	2,666
7	Starwood Hotels	USA	225,014	694
8	Promus Hotel Group	USA	192,043	1,337
9	Carlson Group	USA	106,244	548
10	Patriot American/Wyndham	USA	100,989	472
11	Hilton Hotels	USA	85,000	250
12	Hyatt Hotels	USA	82,224	186
13	Sol Meliá	Spain	65,586	246
14	Hilton Int'l	UK	54,117	170
15	FelCor Lodging	USA	50,000	193
16	Forte Hotels	UK	48,407	249
17	Société du Lourve	France	37,630	601
18	Westmont	USA	37,207	305
19	La Quinta Inns	USA	37,019	287
20	Club Med.	France	36,010	127

[a]Based upon Room Offerings.
Source: *Hotels* (1999a).

(Dunning & McQueen, 1982:73). As a result of mergers, by the late 1980s, several of the largest hotel chains in the world were owned by banks, holding companies, and industrial concerns. For instance, among the holdings as of 1988, ITT owned Sheraton (later selling it off to Starwood), Grand Metropolitan owned Intercontinental, Omni hotels were owned by the Irish airline Aer Lingus, Stouffer hotels by the Swiss company Nestlé, Travelodge/Viscount by Trusthouse Forte (later Forte PLC), Hilton International by Ladbroke, and Holiday Inns International by Bass Breweries (EIU, 1988, 29). As of 1999 Bass PLC owned not only Holiday Inn, but also Intercontinental, Crowne Plaza and Express by Holiday Inn, becoming the second largest hotelier in the world. Starwood Resorts owned two major brands, Sheraton and Weston. Among "discount" hotel chains, Cendant (formerly HFS), the world's largest hotelier, controled the brands Ramada, Days Inn, Howard Johnson, Travelodge and Super 8. It is also integrated into Avis car rental and RCI timeshares.

"Unpackaging" Strategic Assets

Examination of the evolution of hotel chains over the past 40 years reveals one additional feature that is central to the analysis here. Hotels commonly expand into new markets through unpackaging strategic assets. In other words, unlike most manufacturing industries, the above firm-specific advantages may be separated from actual ownership of hotels. The Hyatt Corporation, founded by the Pritzker family in 1957, pioneered the unpackaging of hotels when it created two separate companies to manage the respective real estate and operational aspects of the business (EIU, 1987:46). Others have followed and the result, especially since the 1960s, has been expansion of hotel TNCs largely through alternatives to equity participation (Bull, 1991:189; Dunning & McQueen, 1982:80).

The most common forms of this practice have been entry into new markets through leasing agreements, management contracts, and franchising (Dunning & McQueen, 1982; UNCTC, 1982; Witt, Brooke & Buckley, 1991:64). In leasing agreements chains take over all aspects of the property, including oper- ation, and generally pay the hotel owners a percentage of profit after deducting operating expenses. Franchise agreements vary, but usually include use of the chain's name, trademark and other services such as access to reservation systems in return for a fixed fee along with other percentage-based charges. Technically a franchise is a particular form of licensing agreement that involves a trademark (brand name) and almost always a geographically-based right to sell under that trademark. Franchising may also involve access to more tech- nical expertise that can make it difficult to distinguish from the management contract (Witt, Brooke & Buckley, 1991:70–77). Franchising terms vary signif- icantly. The chain normally provides additional operating expertise, often in the form of manuals and other information, while requiring the individual hotel to maintain certain standards. These can range from accounting procedures to requirements on the frequency of cleaning a swimming pool. They are often enforced by unannounced inspections (*The New York Times*, 1995:D4; UNCTC, 1982:59–60). In management contracts responsibility for various aspects of operation of the hotel fall to the chain itself. Management contracts vary widely, and may range from day-to-day operation to design and construc- tion of a hotel (UNCTC, 1982:60; Sinclair, Alizadeh, Atieno & Adero Onunga, 1992:52–55).

Non-equity forms of participation by chains have grown significantly in recent decades. This is especially the case in expansion into developing countries, where management contracts in particular have become increasingly popular. This favored form of participation may be due to barriers against direct foreign investment or due to political concerns held by TNCs. Frequently, however, nonequity participation is the favored strategy regardless of the political situation. Because, as Witt, Brooke and Buckley point out, hotels have high fixed costs from initial capital costs, but low operating costs, among the biggest advantages for the chain is avoiding high initial capital outlays in construction or purchase of hotels (1991:20–21). Therefore, firms avoid many sunken costs and yet realize considerable benefits from the local market.

The Dunning and McQueen (1982) sample of almost 157,000 rooms in 491 foreign hotels associated with TNC chains confirms growth of non-equity participation by these chains. The percentage of TNC-associated hotel rooms located in developing countries that were owned either in part or completely by the foreign firm, dropped from 21.8 percent to 6.7 percent in the period from 1964 to 1975. After 1975, 90.6 percent of the total number of rooms in that same category were affiliated with TNCs through management contracts (Dunning & McQueen, 1982:81). By the mid-1980s several hotel chains could in fact only be called chains in the sense of associating through these non-equity agreements.

Hyatt Hotels Corp, for example, owns no hotels. It manages more than 150 hotels and resorts worldwide. According to the UNCTC, 100 percent of Best Western's hotels took the form of non-equity (franchised, licensed, or management contract) agreements, while other chains such as Holiday Inn, Ramada, Trusthouse Forte, and Howard Johnson all exceeded 85 percent (UNCTC, 1990:105). The Marriott Corporation expanded aggressively in the 1980s, growing from 75 to 539 hotels, mainly by building hotels to its own specifications, selling them to investors, and then managing the hotels. Ultimately, however, it found itself owning several of the properties and carried more than $3 billion in debt. In 1993 Marriott split into two companies: one owns the real estate (and most of the debt), while the other provides hotel services (*Forbes*, 1995:48–50). For the chains, the terms of the franchising or management agreements are quite lucrative. The terms are generally short in duration, fees paid to the TNC tend to be based on gross receipts rather than profits, and costs associated with required remodeling or redecorating generally fall to the owners (Ascher, 1985; author interviews, 1992, 1995). Major physical alterations usually require owner consent, but many contracts stipulate that this should not be withheld "unreasonably". Chains also usually favor large-scale, capital-intensive luxury hotels. Ascher (1985:44–45) points out that from the standpoint of the chain, hotels that offer less than 300 rooms are seldom profitable, and often chains will push for 600–700 rooms or more. The contracts often contain escape clauses for the TNC (UNCTC, 1982:60–61).

These practices raise the question of control over the specific property, and also the destination as a whole. If, as Lanfant contends, the true nature of TNC power in tourism is that, "... For what they sell is no less than a society, a culture" (1980:23), an additional aspect of that power is the ability to stop selling it. For more immediate purposes, it also indicates that sector attributes shape not only international industrial structures but also local development patterns. This overview is not to predict that the Mexican hotel sector will absolutely mirror its international counterpart; instead the contention is simply that the same set of international incentives will be in place in the Mexican case.

Evolution of the Mexican Hotel Sector

Considerable debate exists over control of hotels in Mexico. Some analysts argue that similar to other Third World countries, the hotel sector in Mexico has fallen into the hands of TNCs (Molinero Molinero, 1982:94–95; Schédler, 1988). Often this forms part of a larger argument that Third World tourism provides few benefits for the local population. In the Mexican case this argument takes two forms. The first holds that foreign chains have become majority owners of many hotels there, thus gaining most of the benefits from tourist stays (Molinero Molinero, 1982). There is little evidence, however, to support this position. A second acknowledges that many hotels are Mexican owned, but TNC-based chains establish control through contractual means. This perspective argues that despite little or no equity holdings by TNCs, their market power deriving from firm-specific advantages has allowed them to wrest actual control of the most dynamic areas of the industry (Mattelart, 1974; Moreno Toscano, 1970; Schédler, 1988). These interpretations are consistent with Dunning and McQueen's (1982) findings, as well as with more explicit dependency approaches to tourism in the Third World (Britton, 1981, 1982, 1991; Lea, 1988; Turner & Ash, 1975).

According to Schédler (1988:138), for example, while only nine percent of Mexican hotel rooms were tied to TNCs in 1987, the picture was very different among international-class establishments. He found that 100 percent of rooms in the luxury "Gran Turismo" category were tied to foreign chains, as were almost 71 percent of Gran Turismo and five-star rooms. The TNCs tend to make decisions regarding overall firm strategy, and those decisions reflect their interests. More recent data confirm ties between local investors and TNC hotel chains, but patterns of ownership and control are more complex than either of these two positions would suggest.

Instead the two key features noted above — reliability embodied in a brand name and the ability to unpackage assets — create obstacles against local firm participation and especially control. They do not preclude it, however, at least among large-scale capitalists in Mexico. Several large and dynamic Mexican firms have become active participants in international class tourists, most commonly through strategic partnerships with TNC-based chains. In short, hotels have not become denationalized, but the distributional patterns of industrial transformation have become narrower as ownership and control has become more centralized.

Early Development Patterns

Due in part to the complexities of ownership and the constantly changing nature of affiliation, data on the structure of hotel ownership in Mexico are at best incomplete. Several trends, however, can be identified. Before 1970, most hotels were independent and had few links to foreigners. Many were small, family run operations. As chains formed and began to expand internationally, however, they quickly established ties in the Mexican market. Inter-continental and Hilton were pioneers, moving into Mexico as early as the 1940s (Cockroft, 1983:152; Schédler, 1988:138). International chains concentrated in the major cities and Acapulco.

As tourism demand grew in Mexico, TNC activity in the hotel sector also expanded. Dunning and McQueen (1982:77) report that by 1978, 39 TNC-affiliated hotels (totaling more than 11,000 rooms) were located in Mexico. This number is not particularly high in itself, given that Mexico had a stock of about 250,000 hotel rooms at that time. Taking into account the segmentation noted in Chapter 4, however, the presence of TNC-linked hotels within international tourist-class hotels was significant. Mexico also ranked first among developing countries in foreign-affiliated hotels and rooms at that time, more than doubling its closest rival, the Philippines. In fact, just three of the 128 developed and developing countries surveyed by the authors had more TNC-affiliated hotels than Mexico and only four had more rooms (Dunning & McQueen, 1982:77).

Despite the lack of complete data, two historical patterns are evident. Early on, during the 1970s and early 1980s, a clear division of labor emerged under which international hotel chains entered the market through means other than equity investment. Meanwhile domestic investors put forth most of the capital to build or buy hotels and then entered into associations with chains. The partnerships tended to be quite fluid and particularly lucrative for the chains, and it was not uncommon for chain and hotelier to part ways after contracts ran out (author interviews, 1992).

The profile of the typical domestic investors in hotel properties during the 1970s remains difficult to pin down as they came from varying backgrounds (personal correspondence with Daniel Hiernaux; author interview, 1992). Some, such as Grupo Alfa, part of the famous Garza Sada family-based Monterrey Group, were among leading national industrialists seeking to diversify holdings, but also common was an array of local professionals and business people who pooled their money to buy a property or two. The state was also active in expanding hotel ownership during this period through the publicly owned Nacional Hotelera. Foreign investment in hotels, however, remained very low. One SECTUR official estimated that more than 90 percent of overall investment in hotels was domestic during this time (author interview, 1992).

International chains continued to pursue this strategy of expansion through non-equity means despite rather extraordinary state efforts to encourage equity investment. Mexico historically placed significant limits on direct foreign investment in the 20th century. Aside from the famous expropriation of petroleum installations under president Lázaro Cárdenas in the 1930s, the state began imposing a series of rules and regulations regarding foreign investment since at least the 1950s (Casar, 1988; Evans & Gereffi, 1982; Fajnzylber & Martínez Tarragó, 1976; Jenkins, 1992; Peres Nuñez, 1990; Whiting, 1992: 70–79). In addition, tourism presented special problems due to constitutional prohibition of foreign ownership of coastal land. The "restricted" zones derive from Article 27 of the 1917 Mexican Constitution and comprise of strips of land within 50 kilometers of the coast and 100 kilometers of international borders.

State actors responded to these problems in 1971 by instituting the framework that legalized a trust mechanism, or *fideicomiso*, in an effort to provide the means to bypass the constitutional restrictions. The new regulations, made by presidential decree in April 1971, allowed the title to land in the prohibited zone to be held in trust for foreign investors by an authorized Mexican bank. The maximum length of the trust was established at 30 years, after which it could either be renewed for an additional 30 years or the property sold to a Mexican citizen. The trusts entitled foreign holders to utilize the property as if they were owners (*Diario Oficial*, 1971; Tancer, 1972; Truett & Truett, 1982:15).

Perhaps the most striking aspect of the trust mechanism for the prohibited zones is its timing. Its unveiling in 1971 took place within a larger environment of growing national suspicion toward foreign investment in Mexico. The tourism trust mechanism was instituted under Luís Echeverría, widely considered the most nationalist Mexican president since Cárdenas (1934–1940). Two years later the Echeverría administration produced the Law to Promote Mexican Investment and to Regulate Foreign Investment, the first economy-wide attempt at controlling foreign investment in the Third World *(Diario Oficial,* 1973). The wide-ranging law codified and expanded prohibitions and limitations on foreign ownership of the Mexican economy. Most important, it placed a ceiling on foreign equity in firms to 49 percent or lower, depending on the area of the economy. The law included a grandfather clause in the form of automatic approval for all existing foreign holdings at their 1973 level. Similarly restrictive laws on technology transfer and inventions and trademarks were also passed during the Echeverría administration (Whiting, 1992:120–130, 248–251). The trusts for coastal zones ran directly counter to these broadly restrictive policies.

Explaining this apparent contradiction in policy toward direct foreign investment requires returning to consideration of the nature and structure of the tourism sector, especially those of hotels. Foreign hotel chains were already entering the Mexican market, and because they favored management contracts and franchising over direct equity ownership, most would easily fit under any equity ceiling. In fact the policy was meant in part to deal with the prevailing division of labor, wherein local investors provided the majority of necessary capital while chains operated the hotels or were affiliated through franchising (de Mateo, 1987:42, 236–237; Moreno Toscano, 1970:251). This arrangement was leading to the situation noted by Dunning and McQueen (1982:70), as well as Schédler (1988), where the local hotel sector was marked by significant foreign control with little or no foreign equity.

From the standpoint of policymakers the choice was clear. If foreign hotel chains were needed to foster exports, and in fact were establishing a significant presence in the hotel sector anyway, they may as well inject some of their own capital into the market. The 1971 trust mechanism that liberalized foreign holdings in the prohibited zones was intended to ease disincentives to equity investment. In the case of the new coastal planned resorts, the state went further. Fearing that the total risk burden would lie with local private investors or the state, FONATUR *required* foreign hotel chains to hold at least 25 percent of the equity for establishments they associated with in Cancún. The requirement was made specifically to counteract TNC entry through non-equity means, most commonly under terms that were, in the words of one public official "… very poor for the owners, and very good for the operators" (author interview, 1992).

Despite these measures, the prevailing division of labor between local investors and TNCs continued through the early 1980s. In Cancún chains initially invested more of their own capital due to demands of state officials, and much of this came in the form of partnerships with FONATUR. Elsewhere, TNCs invested little. This largely resulted from TNC strategies and the assets at their disposal. Most, but not all, TNCs that entered the Mexican market during the period preferred expansion through non-equity means. One large US-based chain, for instance, had earlier established a presence in Mexico with a fully-owned hotel in the capital during the 1960s. According to a company official, the TNC desired ownership of other properties but found the trust mechanism to be lacking in assurances for protecting foreign owners. Firm strategy was also affected by the Mexican government's broader attitude toward direct foreign investment, which was more regulatory in the 1970s. As a result, the TNC expanded in Mexico through non-equity means, which the company official called "very unusual" compared to their other foreign operations (author interview, 1995). Chains were particularly attractive to prospective owners of hotel properties, especially those interested in attracting foreign tourists. Because most of them were from the United States, it was advantageous for hotels to be linked to computer reservation systems, toll-free numbers, and travel agencies there. Most important of all, however, was that a Mexican hotel in a tourism zone be associated with a trade name commonly known in the United States. Simply put, the real estate investors needed the chains more than the chains needed them.

Recent Patterns in Hotels

While investment remained primarily Mexican, then, a growing number of hotels were becoming affiliated with chains, which was especially the case in the larger, high-category hotels in the tourism zones. This began to change, however, where the division of labor blurred and the nationality of the participating firms started to lose meaning. The case of Grupo Posadas (Posadas de México), is illustrative. Posadas is a Mexican hotel and hotel management company that has operated both its own and others' hotels. The company was founded as Promotora Mexicana de Hoteles in 1967 by Mexican businessman Gastón Azcárraga Tamayo. Azcárraga was previously involved in Chrysler de México (Fábricas Automex) and is the first cousin of media baron Emilio (El Tigre) Azcárraga, who until his recent death was principal owner of Televisa. The latter was, in the early 1990s, listed as the wealthiest individual in Latin America by Forbes magazine (Alisau, 1992; Molinero Molinero, 1982; author interview, 1992). The proportion of foreign participation in Posadas itself has varied over the years, but by the 1980s it became a wholly-owned Mexican company. Reflecting the ability to unpackage the hotel product, Posadas first worked with American Hotels (owned by American Airlines) and later some of the hotels Posadas operated but did not own were under franchise agreements with Holiday Inn. In other words, several hotels were owned by one firm, operated by a second, while simultaneously under a franchise agreement with a third. The arrangement became even more complicated in 1976 when Holiday Inn's Mexican subsidiary and Posadas de México merged (Alisau, 1992:12).

Posadas represents an early, and particularly complex example of a second identifiable trend emerging in the Mexican hotel sector beginning in the early and mid-1980s. This trend is marked by three overriding features. First is privatization. The state began to withdraw from ownership in the hospitality sector. Nacional Hotelera, the state-owned hotel chain, held 28 properties when it was privatized in 1985, representing the second largest chain in Mexico at that time. Several were closed upon privatization (SPP, 1985). By 1992 the state continued to own a few assorted money-losing hotels, but most of its holdings had been privatized. As Chapter 4 indicates, it continued planning and to a lesser extent financing, but since the mid-1980s a concerted effort was made toward privatization. Second, the profile of Mexican investors in hotels had changed. Several of the largest Mexican industrial and financial groups aggressively expanded into the hospitality sector, and they began to compete with and replace smaller investors. Finally, while foreign chains maintained a high profile in Mexico, their relationship to the Mexican firms was becoming more strategic, long-term, and complex.

Table 5.2 summarizes the structure of hotel chains in Mexico as of 1992. The top 15, ranked by number of rooms, represented less than one percent of all the hotels in Mexico, but accounted for 25 percent of the hotels and 43 percent of the rooms among those rated in the three top categories. While the list includes a number of familiar TNC-based chains, Mexican firms are also well represented. To some extent, this feature is hidden. The top two chains on the list, Holiday Inn and Fiesta Americana, are both operated by the Mexican firm Grupo Posadas (with the exception of three of the 29 hotels). Posadas, as already above, was long associated with Holiday Inn. The changing relationship between the two companies reflects larger structural changes in the hotel sector. Posadas held the "master franchise" for Holiday Inn in Mexico for many years, and operated several of the hotels under that flag name. The two companies merged in 1976, but Posadas gained operating control in the early 1980s and bought out the Holiday Corp. in 1989 (Alisau, 1992:12; author interview, 1992). In 1992 the company operated 29 hotels in Mexico and eight more abroad, and by 1993 it was expected to add ten more (*El Financiero*, 1992a, 1992b). It held at least some equity interest in 24 of the 37.

By some estimates in 1992 Posadas was associated with about half of the four and five star hotels in Mexico. It fully owned one-third of the properties and

Table 5.2: Hotel Chains in Mexico (1991).

Rank	Hotel Chain[a]	Rooms	Hotels
1	Holiday Inn[b]	4,950	16
2	Fiesta Americana	4,944	13
3	Sheraton	2,818	6
4	Best Western	2,658	21
5	Calinda Clarión	2,544	15
6	Camino Real	2,387	8
7	Stouffer Presidente	2,252	7
8	Vista	2,066	7
9	Melia	2,034	7
10	Exelaris Hyatt	1,862	6
11	Misión Park Inn	1,752	12
12	Princess	1,363	2
13	Krystal	1,304	4
14	Paraíso Radisson	1,285	5
15	Plaza Las Glorias	1,173	8

[a]The chains are listed by trade names. At times the same firm will operate or franchise hotels under different names. List does not include all-inclusive chains.
[b]Includes Holiday Inn and Holiday Inn Crowne Plaza name, and includes three hotels not operated by Grupo Posadas.
Source: PKF (1991).

held a minority stake in another third (Alisau, 1992). Other estimates are lower, indicating Posadas' association with 27 percent of the five-star and Gran Turismo rooms, seven percent of four-star rooms, and 23 percent of the three categories combined. This still placed Posadas as easily controlling the largest portion of the combined market (Internal market study, private sector firm, March 1992). Grupo Posadas diversified as well, moving into more lucrative areas of hotel operation previously dominated by foreign firms. It continued to own properties, and to operate hotels under the Holiday Inn, Holiday Inn Crowne Plaza, and Fiesta Americana names, but also started a new franchise trade name, Fiesta Inn. It diversified investment partners, developing new properties with Swiss, German, and other Mexican investors in the 1990s. In addition, Posadas initiated a tour operator subsidiary in an effort toward vertical integration. Posadas also became a TNC itself as part of aggressive expansion during the 1990s, opening new hotels in Venezuela, Texas, and California. In the early and mid 1990s it opened a new hotel every two months. By 1995 Posadas was associated with nearly 8,000 rooms and planned expansion of more than 1,000 more in the next three years, making it the largest provider of hotel rooms in Latin America. Those figures only include rooms under the Fiesta Americana and Fiesta Inn trade names (*Hotels*, 1995:38). By 1998 it ranked third in hotel rooms in Latin America behind Bass (Holiday Inn) and Starwood Hotels (Sheraton and Weston) with just under 10,000 rooms in four countries (*Hotels*, 1998) and is more than three times as large as its nearest competitor in Mexico (*Hotels*, 1999b). Posadas acquired the Brazilian brand Caesar's Park in 1998 and is active there and in Argentina.

Tables 5.3 and 5.5 demonstrate the leading role played by Posadas and also show that the Mexican hotel sector has not become denationalized. Several large Mexican business groups have moved into the hospitality sector, most commonly by developing longer-term strategic alliances with international hotel firms. Among the Mexican investors entering the hotel business during the 1980s were industrial conglomerates grupos ICA, Cemex, Gutsa, and Sidek, along with the large domestic banks Banamex and Bancomer (author interviews, 1992; Morgan Stanley, nd).

Ingenieros Civiles Asociados (ICA), the largest engineering and construction firm in Mexico, provides a second useful example. The conglomerate, made up of more than 180 separate companies, has been a consistent beneficiary of government contracts throughout its 50-year history. ICA was founded in 1947 and has had a hand in virtually every major Mexican public project since 1950, including the new sporting venues for the 1968 Olympics, the expansive national university campus (*Ciudad Universitaria*), the Mexico City subway, and public housing installations at Tlaltelolco, along with several major highways and ports. By 1980 ICA was the largest construction firm in Mexico and is

Table 5.3: The Largest Hotel Chains in Mexico (1995).

Firm	Country	Rooms	Hotels
Grupo Posadas	Mexico	8,493	34
Holiday Inn Worldwide	USA	6,159	35
Grupo Situr	Mexico	6,135	25
Hoteles Camino Real	Mexico	3,556	15
Best Western International	USA	3,281	36

Source: Hotels (1996).

now considered the largest in Latin America. It has become a TNC, undertaking projects throughout Latin America, and working with US-based companies such as Waste Management and Bechtel (*El Financiero*, 1992c; Munguía Huato, 1989).

Included among ICA's projects in Mexico were several in tourism infrastructure. Like other recipients of the public works in the industry, the company eventually became a provider of tourism services. According to one official, ICA received the contract to build the major highway spanning the Baja California peninsula (Highway No. 1). As part payment, it received title to land in the coastal development of Loreto/Nopolo/Puerto Escondido. ICA then traded that land for hotel sites in other tourism poles (author interview, 1992). ICA entered mainly in hotel real estate, eventually forming partnerships with Sheraton and Radisson. By 1990 it wholly or jointly owned several hotel properties along with management companies to run the hotels under the Radisson or Sheraton names. The partnerships, in fact, became quite complex. ICA formed a hotel promotion firm called Paraíso in 1986, in an equal partnership with Banamex, Mexico's largest private bank. The new company invested in five luxury hotel properties, in most cases as a minority equity partner that also managed the hotels. Meanwhile it also reached contracting agreements with three Radisson-associated companies that gave it trademark rights, access to computer reservation systems and technical assistance. ICA had also teamed up with Sheraton since 1981, becoming the majority owner of Sheraton's Mexican subsidiary, and operating six hotels in the "Gran Turismo" category. Sheraton moved beyond simple franchising and technical assistance agreements, however, to also invest in the properties. Meanwhile, ICA, like Posadas, also moved beyond simply providing investment capital by moving into some aspects of the more lucrative operating area. Since 1992 ICA has begun to limit its hotel business expansion and it recently sold its hotel operating subsidiary (author interviews, 1992, 1995; ICA, nd).

Grupos Sidek and Cemex (*Cementos Mexicanos*) also moved from industry and construction into tourism development. Sidek, primarily a steel producing company based in Guadalajara, formed a tourism division (Situr) that aggressively moved into resort development and hotel operations. In hotels Situr teamed up with Posadas and Sheraton to hold interest in more than a dozen hotels, operating under the Continental Plaza, Plaza las Glorias, Fiesta Americana and Sheraton names (Morgan Stanley, nd). More recently it also owned and operated Holiday Inns. In addition, Situr successfully bid for the concession to develop the FONATUR Megaproject Marina Ixtapa, along with additional tourism and real estate developments in Baja California, Puerto Vallarta, and Mazatlán (*Business Mexico*, 1992; *El Financiero*, 1992e; SECTUR, 1991:8).

Situr also purchased some hotels previously associated with Radisson and ICA/ BANAMEX in 1993 (author interview, 1995). By 1994 Situr had aggressively expanded, and was associated with at least 25 hotel properties. The company was severely affected, however, by the 1994/1995 peso crisis as much of its debt was dollar denominated. It teetered on bankruptcy before having much of its debt restructured. In the process it lost many of its hotel holdings.

Meanwhile Cemex, the third largest cement producer in the world, formed an alliance with Marriott. Cemex, like Sidek, moved into the hospitality sector through strategic alliances with TNCs. Other large Mexican industrial and financial groups have followed suit. The banks Banamex and Bancomer worked with Radisson/Inter-Continental and Hilton, respectively, although the former later was associated with Presidente. In other alliances, Hoteles Presidente, the privatized hotel group once known as Nacional Hotelera, teamed with Stouffer, Grupo Xabre, one of the primary holders of Mexicana Airlines, with Westin, with Grupo Carso, which holds controlling voting interest in the telecommunications giant Telmex (*Teléfonos de México*), and with Choice, operators of Quality Inns (Calinda). The flag names of the latter joint venture include Quality Inn, Calinda, Clarion Calinda, Comfort Inn and Clarion. They included 17 different ventures as of 1992, mainly in cities. Carso is headed by Carlos Slim Helú, who according to Forbes was worth $6.6 billion in 1994. Slim is cousin to Alfredo Harp Helú, a founder of Banacci, Mexico's largest financial group and owner of Banamex. Table 5.4 summarizes the primary actors and alliances within Mexican hotel chains in the early 1990s. In addition, other large conglomerates such as DESC, which gained the concession to Punta Ixtapa, have invested in tourism real estate through the Megaprojects.

There data contradict suggestions that the Mexican hotel sector has become denationalized. During the period in question, TNCs have stepped up their presence in the Mexican market. Meanwhile, their national partners tend to be among the more powerful, internationally-oriented business groups. Each is represented in *Expansión 500*, the Mexican equivalent to the Fortune 500, and most trade on the New York Stock Exchange. According to the 1992 version of *Expansión* (1992), titled "The Most Important Mexican Firms," TELMEX (controlled by Grupo Carso) was ranked 2nd, Mexicana Airlines (controlling interest held by Xabre) was ranked 9th, ICA 25th, CEMEX 30th, Situr 65th, and Posadas de México 235th.

It is important to note that these rankings were only for individual firms, and often do not reflect the size of the overall holdings of a conglomerate. ICA, for example, was listed 25th, but this reflects only its core construction business. ICATUR, ICA's tourism division, was also on the list, ranked 207th. Desc, a holding company, had interest in nine separate firms, eight of them joint ventures

Table 5.4: Strategic Alliances in the Mexican Hotel Sector (1992).

Firms	Flag	Hotels	Rooms
Posadas/Holiday Corp.	Fiesta Americana	13	4,975
	Holiday Inn	9	2,487
	Holiday Inn Crowne Plaza	3	1,309
	Fiesta Inn	4	457
	(various others)[a]	7	1,431
Situr/Posadas/Sheraton	Continental Plaza	13	3,902
	Plaza las Glorias		
	Fiesta Americana		
	Sheraton		
Carso/Choice	Calinda Quality Inn	18	3,002
ICA/Banamex/Radisson	Paraíso Radisson	4	1,176
Banamex/Inter-Continental	Sierra	3	847
ICA/Sheraton	Sheraton	6	1,741
	Pirámides del Rey		
Xabre/Westin	Camino Real	8	1,108
Bancomer/Hilton	Conrad Hotels	3	n/d
Presidente/Stouffer	Stouffer Presidente	7	2,170
Cemex/Marriott	Marriott	2	883
	Marriott Courtyard		

[a] Refers to management and ownership interest in seven hotels in the United States operating under the brands Holiday Inn, Sheraton Fiesta, Border Inn, and Hampton Inn.
Source: Compiled from company reports, internal private sector market study, March 1992, undated, untitled overview of lodging companies in Mexico produced by Morgan Stanley, newspaper reports, author interviews, Mexico City, 1992.

with TNCs. Bancomer and Banamex were ranked first and second in the separate banking list. All rankings are in terms of 1991 annual sales, or in the case of banks, assets. In going beyond individual firms to examine business *grupos* or conglomerates, Roderic Camp (1989:177–189, 200–205) includes ICA,

CEMEX, and DESC among his list of the 50 most powerful business groups in Mexico. Matilde Luna (1992:110), a leading scholar of Mexican entrepreneurs, includes ICA, Bancomer and Banamex among the country's top eleven business groups. *Forbes* (1994) listed all the primary families associated with Banamex, Carso, Cemex, and Sidek; individuals from those families and firms were among the world's wealthiest.

The peso crisis of 1994–1995 had a profound impact on the hotel sector in Mexico. The most immediate result was a halt on construction. Several firms also found themselves heavily indebted, most notably Situr and Camino Real hotels. Several properties that were managed or franchised also went bankrupt. Only in 1998 and 1999 did investment return in any significant manner to the hotel sector. This has also led to a shakeup in the leading players in the hotel sector in Mexico, as Table 5.5 demonstrates. The overall patterns of a mixture of Mexican and international capital continues, however. One of the newest players, Grupo Chartwell, is indicative of this. Chartwell, founded in 1995, is a globalized company. Although based in Mexico City it contains private investors from both the United States and Mexico and is backed by both Mexican steel interests as well as a large foreign institutional investor in Credit Suisse First Boston. It has grown rapidly and also recently struck up a strategic franchising alliance with US-based Hilton Hotels (*Hotels*, 1999b).

Another relatively recent change in the hospitality sector is that of the source of capital brought into hotels. Increasingly foreign chains and other foreign investors have begun to inject capital into hotels and other related tourism projects. In part this resulted from attractive debt-for-equity swaps offered by the state in the late 1980s. Data from the Ministry of Trade and Industry, reported in SECTUR (1991:353) and *LatinFinance* (1990:30–32), show that between 1986 and 1990 the government approved swaps involving more than $2.8 billion.

Table 5.5: Mexico's Fastest Growing Chains (1999).

Firm	Country	Rooms	Hotels	Rooms to Open, 1998–2001
Posadas	Mexico	9,078	40	2,367
Bass Hotels	USA	6,571	24	1,710
Cendant	USA	2,941	28	1,440
Occidental	Spain	1,112	2	1,000
Chartwell	Mexico	2,200	9	772

Source: *Hotels* (1999b).

Tourism was among the most popular industries, accounting for almost 39 percent of the value of all the swaps during the period. Many of these swaps categorized under direct foreign investment in fact constituted repatriation of domestic flight capital (author interview, 1992). Although the swaps were discontinued after 1990, tourism continued to attract much higher levels of foreign investment than previously. In part this was likely due to the further liberalization of the trust mechanisms in the prohibited zones in 1989. Under the new provisions the 30-year term of the trust would be automatically renewed upon request (FONATUR/SECTUR nd). While the figures provided in the tourism category are not broken down, a significant share appears to be directed at hotels. This process continued throughout the NAFTA negotiations and ratification — partly in anticipation of expected increases in business travel — and only began to slow with the peso crisis in 1994–1995 (author interview, 1995).

Conclusion: Changes in the Mexican Hotel Sector

The hotel sector in Mexico experienced a steady transformation over the past 25–30 years. Among the higher categories of hotels, a few TNCs entered the Mexican market shortly after World War II, not long after they themselves were formed. Most were confined, however, to Mexico City and Acapulco. As the state began to promote tourism exports after 1968, however, foreign firms steadily moved into Mexico. Most often this was initially accomplished through contractual means regarding brand names, technical assistance, and operating agreements. State efforts encouraging foreign hoteliers to inject capital into the market met with little success during the 1970s and early 1980s, and TNCs, bound only by contracts, floated between hotel properties.

As of late, however, this pattern has changed. The upper echelon of the Mexican hospitality sector is now marked by a series of strategic alliances between internationally-based hotel operators and large Mexican business groups. To be sure, these alliances vary significantly. Posadas, ICA, Situr, and others have moved from real estate investments into the lucrative and less risky hotel operating business. By the early 1990s subsidiaries of Posadas and ICA, for example, operated but had no equity interest in at least some hotels. Situr started its own brand name. Posadas has also developed its own state-of-the-art computer reservation system as well as support subsidiaries involved in hotel and restaurant supply (author interview, 1992). Others such as Cemex, however, remained largely passive investors, while partner Marriott International took on responsibility for hotel operations (author interview, 1992). In addition to these variations, the strategic alliances themselves are not set in stone. Posadas and Holiday Inn steadily began moving apart beginning in the late 1980s. Situr invested in several Fiesta Americana hotels with Posadas at one time, but later teamed with Holiday Inn with some properties. Since 1992 Inter-Continental and Banamex split, the former moving to associate with Presidente. Bancomer and Conrad Hilton also parted ways in late 1992 as the two big banks began to move out of tourism as they were to be privatized.

Despite these recent changes in partnerships several conclusions are apparent. First, expanding the Mexican hotel sector has not been accompanied by denationalization. Instead the market has become more segmented and at the same time centralized, with top categories of hotels becoming tied to international *and* the most dynamic fraction of domestic capital. Second, if, during the 1960s and 1970s, the predominant pattern was a division of labor between foreign hotel operators and domestic investors, the relationship has blurred in recent years. Some local firms are becoming more involved in hotel management, franchising, and support services. Posadas has gone the furthest, but others such as Situr and Xabre have moved into management roles. One indication of the blurring of roles was the attempt by a group of Mexican investors to purchase the Westin Hotels and Resorts Chain from the Tokyo-based Aoki holding company in 1994. The bid, on what represented the 17th largest chain in the world at the time, however, failed (*Hotels*, 1994:3). Meanwhile, foreign investment in tourism continues. While some of this constituted portfolio investment in Mexican firms involved tourism, especially in 1993 and 1994, it also involves significant direct equity investment in properties.

Third, these distributional patterns and roles are best understood by paying attention to industry attributes which in this case derive from the hospitality product itself, and create firm-specific advantages that shape the manner in which local development of the sector emerges. For hotels two factors — trust or reliability and the ability to unpackage assets from ownership — are the crucial aspects that shape sectoral development. While recent trends in the Mexican market indicate that strategic advantages held by hotel TNCs — including operational expertise, technological advantages, name recognition, and trust — are not set in stone, they appear at the very least to be difficult to overcome. Only the most dynamic fraction of Mexican business has been able to gain access into the more lucrative operation and franchising aspects of the hospitality sector. Indeed international chains only seem interested now in association with local firms that are large, diversified and with access to substantial credit lines. Even then, many Mexican firms continue to maintain ties to the international chains. In short, industry attributes have largely assured that the distributional gains from the tourism push, at least in hotels, would be narrow.

Could it have been otherwise? Certainly, a more equitable outcome was at best unlikely. Altering existing distributional patterns would require the state's willingness to redirect benefits in some manner, most likely through bargaining. But state managers had much at stake in the chains' presence. Moreover the state had, however unwittingly, advanced these alliances in a few ways. First, and most important, it had built the export push around mass-based resort tourism. This meant it needed luxury hotels for success, and that all but ensured a major role for TNC-based chains. The state also provided the infrastructure and reduced costs and risks for the private sector. Both served to attract large Mexican firms. Several were recipients of state-financed tourism projects, either receiving public contracts for building infrastructure or, in the case of banks, having state-subsidized loans channeled through their institutions. Both Banamex and Bancomer, for instance, were among the top intermediary institutions over the years for loans subsidized and guaranteed by FONATUR. Many of these large private firms then became active participants in hotels, and also became recipients of these loans. As Chapter 4 documented, a greater share of state financing for hotels has gone into the luxury categories in recent years, categories dominated by the same hotels listed in Tables 5.2 and 5.3.

In fact state attempts to redistribute benefits through bargaining were absent in the hotel sector. While it could be argued that this outcome actually reflects state interests — meaning that the tourism push was intended to primarily benefit capital — a more nuanced interpretation is warranted. By the 1990s state actors came to face increasingly powerful counterparts in the hotel sector in the form of TNCs that had few sunken costs and that formed alliances with firms making up the most powerful fractions of Mexican capital. As a result, the question was not one of state willingness to redistribute tourism benefits, but rather one of state autonomy itself, at least in the hotel sector.

Chapter 6

Patterns of Airline Development

The previous chapter details growth and distributional changes of hotels in Mexico over the past 30 years. It demonstrates that prevailing patterns during that time period are primarily a product of sector-specific features derived from actual production and consumption of hospitality services, mediated by state policies and international norms and agreements. These in turn create international structures which influence local development. As a result, today ownership and control of tourist-class hotels in Mexico are largely confined to TNC-based chains and internationally-oriented domestic capital.

This chapter continues to examine subsectors of international and Mexican tourism by turning to air transport. Airlines, along with hotels, form the primary subsectors of tourism and account for the bulk of tourism expenditure. Although technically a part of the transportation sector — and listed as such under balance of payments statistics — airlines have become indispensable to the growth of global tourism. Most tourists who travel internationally do so via air. In many cases today the air transport component of a vacation may make up half or more of total consumer expenditure. Air transport has also become central to Mexican tourism exports, with more than two of three international tourists to Mexico in 1990 arriving by air (SECTUR, 1992).

The chapter proceeds in a manner similar to the previous one on hotels in that it traces the evolving structure of the airline sector in order to uncover distributional aspects of industrial transformation and their determinants. In other words, it discusses industry attributes and international industrial structures, examining their impact on changes in ownership and control that have taken place in Mexican airlines. If sector-specific attributes are as important as the model claims, one should expect at least some variation among tourism subindustries.

At issue is also whether state action can alter the distributional aspects of local development patterns. In other words, is the causal arrow between international and domestic industrial structures absolute and unmediated? This is taken up in Chapter 2 and at the end of Chapter 5 where it is suggested that states holding the necessary technical capacity and material resources may step in to alter prevailing patterns. The costs and benefits of such actions, however, must be taken into account. In hotels the costs were simply too high. State officials needed international hotel chains in order to attract tourists. While costs have been high in air transport, they were also accompanied by significant real and perceived benefits due in part to airlines' strategic status. Even though Mexican air transport has gradually been liberalized as of late, this chapter demonstrates that it continues to be much more heavily regulated and protected by the state than either the hotel or tour operator business. In fact the history of Mexican air transport is filled with episodes of state intervention.

Scheduled Air Carriers

International civil air transport is relatively young, and the business of airlines evolved in a unique manner during the last century. It is the product of market forces similar to that of other services but contains a strategic twist due to the nature of the activity itself. In terms of industry characteristics, air transport resembles other technologically sophisticated and capital intensive sectors. Both of these factors create high entry barriers. Startup costs are high due to the need for expensive equipment and a skilled labor force. In addition production is fairly inflexible in the short term, although it may be adjusted in the medium and long term. As a result, the sector is periodically plagued by problems of excess capacity. Economies of scale exist, but fixed costs are also high and tend to contain a cyclical spike reflecting the cost of updating equipment (O'Connor, 1989; Petzinger, 1995). Further, the nature of production means that all commercial passenger aviation is regulated in some way, if only for scheduling, air traffic control, maintaining take-off and landing slots, gates at airports, and general safety. All of these factors suggest that oligopolies are likely to occur and in fact they do. Most individual markets — that is routes — are served by just a few carriers. Depending on the number of total routes, frequency, and overall demand, a given national and even regional market may exhibit a similar profile of few producers.

A small number of beneficiaries in the scheduled air carrier business is then primarily explained by these basic attributes. What they do not fully explain, however, is the particular international industrial structure of airlines as it has evolved historically. In fact, because of high costs combined with the unique and strategic nature of air transport, firms have been the subject of tight domestic and international controls that have produced significant amounts of state ownership, simultaneous national oligopolies, and controlled international competition.

International air transport is uniquely strategic in several ways. Whether carrying passengers, mail, or other cargo, commercial airplanes not only reach borders (as in shipping) but penetrate what has become recognized to be sovereign air space controlled by nation-states. Because national defense routinely involves the monitoring of air space, governments require that commercial air patterns be easily identifiable and regularized. National defense concerns have also led governments to pay particular attention to this means of transportation. Governments tend to favor developing some type of national carrier(s), in part due to the added reserve capacity for moving troops and materiel in times of defense needs (which the United States did in preparation for the Gulf War), and in part for the spillover effects. The movement of considerable numbers of people also involves safety concerns and invites international and domestic regulation. Further, the capital and technologically intensive nature of the industry invites government attention because of the tendency toward oligopoly. This can be seen when looking beyond general market share statistics. For instance while a half dozen major carriers and various smaller or regional ones serve the US domestic market, the market between city pairs is frequently dominated by one or two airlines and certain "hubs" tend to experience near monopoly conditions (Petzinger, 1995; Morrison & Winston, 1995).

All of these factors have played a role in producing the rather unusual development of the international airline sector during the last 50 years. Airlines are a capital-intensive venture, and as such ownership has been concentrated among fairly large-scale business groups or the state itself. Yet the sector today is also the product of wide-ranging domestic and international government action. As a result, nearly every country has domestic airlines of its own and international commercial air transport is as much the product of state-to-state bargaining as it is of market conditions. Commercial airlines must seek special permission resulting from government negotiations in order to serve a foreign country. Investment historically has been even more limited. Unlike most industries where TNCs may enter local markets for the purpose of serving them, this practice is all but prohibited in commercial air service. These two factors make airlines unique compared not only to hotels, but also with respect to most other economic activities.

Tight regulation has been the norm for international air transport since its inception early this century. Currently it is governed by a system of rules and norms created at the Chicago Convention in 1944 (Doganis, 1993; Jönsson, 1981). This was called primarily in order to find a way to govern the growing air transport sector and produced the so-called third and fourth freedom rights that today allow for international air traffic (OECD, 1993:91). The existing system also reflects the fact that in addition to carrying out business, many airline companies in effect serve the role of showing the flag as a national carrier (Jönsson, 1981).

The two most significant regulations emerging from Chicago were (and are) cabotage and bilateral air agreements. Cabotage, which existed earlier but was legitimized at the Chicago Convention, refers to the prohibition of foreign carriers serving solely domestic routes. Pure cabotage has been chipped away at since, primarily through continuation services, which allow point to point service within a country as long as the flight originates or terminates in the home country of the carrier. Nevertheless, cabotage continues today and forms the economic basis for the existence of domestic airlines in most countries in the world. Bilaterals, also known as Air Service Agreements (ASAs), refer to the two-country negotiated accords that govern all air transport between them.

Bilaterals emerged from Article I of the Chicago Convention, which held that air space above a country is sovereign territory and thus requires state authorization for foreign flights. The accords were clearly a "second best" option after the failure to reach a multilateral agreement at Chicago. The convention was marked by what Eugene Sochor (1991) calls "aviation competition" especially between the two leading nations in terms of airlines, the United States and the United Kingdom. The former, which had strong domestic private airlines, favored an open skies policy that minimized regulation of foreign carriers. Britain called for an international regulatory body with broad powers in setting standards, routes, and prices. In the end neither got what it wanted. Open skies was defeated and although the conference produced the International Civil Aviation Organization — a UN body — it has almost no enforcement power (Doganis, 1993; Golich, 1990; Jönsson, 1981; Sochor, 1991).

Bilaterals, which remain in place today, set the routes, frequencies, capacities and fares for airlines traveling between any two countries. Without a bilateral agreement, no commercial air transport takes place between them. One of the most important early bilaterals was made between the United States and the United Kingdom, stemming from negotiations held in Bermuda in 1946. Other similar agreements, known as Bermuda-types, have followed and are characterized by being more liberal than other bilaterals in that they leave many of the details over flight frequency and passenger capacity to the principal airlines involved, subject to government approval (Sochor, 1991:15–16; UNCTC, 1982). Similar, the even more liberal agreements today are commonly referred to as "open skies" agreements. Even though air transport has grown phenomenally since 1944, bilaterals continue to govern air transport, and more than 1,800 exist today (Findlay, 1990:77).

While the Chicago Convention and the resulting practice of bilateral ASAs distinguish air transport from many other services, three other developments since then have also heavily influenced the sector. First, common government practice severely limits foreign investment in domestic airlines. Again this results mainly from national security concerns and a desire to show the flag abroad. Second, the International Air Transport Association (IATA) — an airline trade association — was formed by 82 carriers shortly after the close of the Chicago Convention, and quickly gained the ability to set international fares (Sochor, 1991:13). This was especially the case in Bermuda-type agreements (Golich, 1990:158–160). As a result, pricing for international flights followed a strict regulatory regimen. Fares were set by IATA members flying specific routes, which had themselves been determined by individual governments, stemming from the terms negotiated in ASAs. Frequently airlines agreed to share revenues on routes, as well as other aspects of providing service, (such as ground crews, catering, etc.).

This classic cartel arrangement lasted for 30 years and only began to break down due to the final development: deregulation. The deregulation of the US market in 1978 brought with it pressure for international liberalization (Vietor, 1994). The US government, representing several of the strongest private carriers and holding the key to the largest domestic airline market in the world, used both assets to renegotiate several bilaterals on more liberal terms. Between 1977 and 1980 the United States successfully renegotiated bilaterals with 15 countries, and several more were completed by 1985 (Doganis, 1993:52–54; Sochor, 1991:35–37). Other nations subsequently followed suit (Johnson, 1993). Gradually the ability of the IATA to set international rates deteriorated as a result of competitive pressure (Feldman, 1987; Golich, 1990; Sochor, 1991:208). The United States played a central role in the demise of the IATA cartel, threatening to subject it to domestic antitrust laws and then to withdraw (de Murias, 1989; Golich, 1990). In the last decade liberalization has accelerated throughout the globe, with the United States taking the lead through pursuing new bilateral "open skies" agreements with several key European and Western Hemisphere nations. Within the European Union, deregulation of markets and privatization of firms has gained speed since 1986 when the European Court ruled air transport to fall under the competition laws set by the Treaty of Rome (Golich, 1990:165–166; Johnson, 1993:222–223).

The impact of the post-war bilateral system is subject to some debate. Unlike other transportation sectors such as shipping, for instance, open registries or flying "flags of convenience" are not an option and one likely result has been that the safety record of air travel has been comparatively strong. Yet by lying outside the General Agreement on Tariffs and Trade (GATT) and later the World Trade Organization (WTO), air transport has also been among the most protected trades in the world (Doganis, 1993; Findlay, 1990; Golich, 1990; Krasner, 1985; Sochor, 1991). Therefore, individual airlines are the product of the 50-year old system set up at Chicago as well as more recent regulatory changes. As Golich and others point out, on the one hand they are international businesses that form strategies based on profit maximization. On the other, however, ownership often includes public participation, operations take place in a heavily regulated market, and at times airlines become embroiled in foreign policy disputes, such as when the United States government prohibited air travel between it and Haiti as part of its larger effort to restore ousted president Jean Bertrand Aristide.

The distributional effects of this system of governance, which are of primary concern here, have been rather straightforward. On the one hand, the capital and technological-intensive nature of the industry all but dictate that the structure of ownership will be concentrated among a few highly capitalized firms. In fact, capital requirements and risk are so high that public participation is frequently deemed necessary. On the other hand, although these same attributes would also suggest that AIC-based carriers would predominate, they do not. Due mainly to the strategic nature of the industry — and its broad recognition as such by most governments — most countries possess their own carriers where airline TNCs are precluded from the domestic market and have a less than overwhelming share of international traffic. Both result from what Krasner refers to authoritative allocation rather than market-based allocation and are embodied in the Chicago Convention and the subsequent regime of ASAs (1985: 196–198, 201–203).

Recent deregulation and privatization, however, have produced important changes. These include growing firm concentration within domestic and some regional markets as many carriers have merged or gone bankrupt. In the United States, five carriers had 71 percent of the market share and four earned 90 percent of the profits in 1989. The market in Europe is also dominated by a small number of carriers (Bull, 1991; Johnson, 1993). Entry costs, usually quite high, have been reduced in recent years due to surplus of planes and a growing practice of leasing. Competition, however, has resulted in most new carriers either folding or merging with larger airlines (Lundberg, Stavenga & Krishnamoorthy, 1995:102–103). In addition, while laws and regulations continue to limit cabotage rights for international carriers and majority foreign ownership in most countries, airlines have pursued alternative strategies abroad. As Chapter 5 indicates, many have integrated with other tourism activities. Many international carriers, such as SAS (Inter-Continental), Air France (Meridien), KLM (Golden Tulip), and Japan Airlines (Nikko) have also become involved in hotels. The German airline Lufthansa has established charter and tour operator companies in its home market and elsewhere in Europe (Bull, 1991; Feldman, 1987; Vietor, 1994).

Since the 1980s several airlines have also pursued a series of global strategic alliances that include cross investment and flight coordination through code sharing. The latter links passengers flying on more than one airline. The practice has grown among domestic feeder airlines with major carriers in the United States recently, but it also may link separate airlines in different countries. A passenger buying a Royal Dutch KLM ticket to South Bend, Indiana, from Rotterdam, for example, may be channeled into Northwest's hub at Detroit and then flown Northwest on the final leg, where KLM has no landing rights. The ticket itself, however, may show the passenger traveling on KLM throughout. Code sharing is a response to both limitations on foreign investment and cabotage. In the United States, for instance, foreign investment in airlines is limited to a 25 percent equity stake since the 1958 Federal Aviation Act.

Finally, many compete through computer reservation systems (CRS), which display routes and fares for travel agents and others who book travel. They also carry information on hotels, car rentals, and other tourist-related services. When a travel agent books a seat for a flight on one airline through a CRS owned by another, the latter collects a fee from the former. Whether CRS expansion is a means of cooperation or competition is a question of some debate but it clearly is a major source of gaining market share and profit for carriers. By 1988 the US Department of Transportation calculated that domestic CRSs earned profits of more than $1 billion (Lundberg, Stavenga & Krishnamoorthy, 1995). AMR, for example, the parent of American Airlines, also owned SABRE, the most widely utilized CRS in the world, until spinning it off in 1999. For much of the 1990s SABRE was the most profitable division in the company. Airlines also aggressively market CRS systems. In 1994, for instance, SABRE signed a 20-year contract with Canadian Airlines valued at $72 million. SABRE provides CRS services to 350 airlines worldwide, the English Channel Tunnel, and is selling a version to Holiday Inn Worldwide for its hotel reservations (*Aviation Week and Space Technology*, 1994; Lundberg, Stavenga & Krishnamoorthy, 1995).

To the extent that airlines have become more global in scope — that is, not just adding more international destinations but rather in cross-investment, horizontal, and vertical integration and licensing or selling of technology — the most dominant emerging airlines in recent years are based in AICs. In 1998 all of the top ten and 17 of the 20 largest carriers were based in AICs, as Table 6.1 demonstrates. As deregulation continues globally, airlines have begun to look more and more like other TNCs.

Table 6.1: World's Largest Scheduled Airlines (1998).[a]

Rank	Airline	Country	Millions
1	United Airlines	USA	200,421
2	American Airlines	USA	175,249
3	Delta Airlines	USA	166,154
4	British Airways	UK	116,101
5	Northwest Airlines	USA	107,402
6	Continental Airlines	USA	79,778
7	Japan Airlines	Japan	78,813
8	Lufthansa	Germany	75,438
9	Air France	France	74,598
10	U.S. Airways	USA	66,389
11	Singapore Airlines	Singapore	57,737
12	KLM	Netherlands	57,279
13	Qantas	Australia	56,675
14	All Nippon Airways	Japan	53,713
15	Cathay Pacific	China (Hong Kong)	40,654
16	TWA	USA	39,279
17	Air Canada	Canada	37,240
18	Alitalia	Italy	35,561
19	Thai Airways	Thailand	34,340
20	Iberia	Spain	32,496
44	Aeromexico	Mexico	10,720
50	Mexicana	Mexico	10,319

[a]Scheduled Passenger Kilometers Flown, Domestic and International.
Source: IATA (1999).

Evolution of the Mexican Commercial Airlines

Mexico, like other Third World countries, has been significantly influenced by the two defining trends of air transport noted above: profit maximization by individual firms and heavy government regulation of domestic and international markets. Not surprisingly the Mexican market is both segmented and marked by limited competition. By the early 1990s the two largest Mexican airlines, Mexicana (*Compañía Mexicana de Aviación*, or CMA) and Aeroméxico, controlled more than three quarters of the domestic market and about half the international market to Mexico. Mexico is somewhat unusual among Third World countries in that it possesses two national carriers rather than just one. Despite this, they are each the product of cabotage rules and ASA's, an imperfect capitalist market, and especially repeated state intervention. During the 35-year time period in question, important changes also took place within the airline sector and both companies. The Mexican national carriers, which had a history of public ownership, were privatized during the late 1980s, but the state engineered a bailout less than a decade later following the 1994–1995 peso crisis. In the case of foreign airlines, the regulatory environment was gradually eased and more carriers entered what was becoming a lucrative market. Despite these trends (detailed below), the evolution of the airline sector in Mexico has followed a logic different from that of hotels. It is, however, the product of sector attributes: high capital requirements and risk combined with its strategic status invited state participation at an early stage in the form of partial and later complete ownership.

Historical Trends

Commercial air service within Mexico can be traced back to 1924 when two US citizens formed CMA (Davies, 1964:76–77). The carrier was started by Americans in order to carry payrolls to oil fields near Tampico. Five years later Pan American, which began serving Latin America to provide mail service and came to dominate the region's early air travel, bought CMA outright (Burden, 1943; Davies, 1964; Leary, 1992). Pan Am also began absorbing new regional entrants into the Mexican market. One of the small carriers Pan Am established interest in was Aeronaves de México (forerunner to Aeroméxico), which first began operation in 1934 with a route between Mexico City and Acapulco (IMIT, 1980). Pan Am bought 40 percent of the airline in 1940. Meanwhile, the Mexican government, like many in Latin America, had started to impose requirements on the private airlines, most notably by regulating individual routes in 1932 through a practice of granting 30-year concessions and by mandating that 90 percent of employees be Mexican nationals (Burden, 1943; Hanson, 1994). This marked the beginning of close state supervision over the airline sector in Mexico. Over the next four decades the two primary concerns of state officials were to maintain a national carrier presence and to protect and maintain that presence both domestically and internationally. Each of these goals was pursued time and again through domestic regulation and investment, on the one hand, and international negotiation of bilaterals on the other.

After World War II, Mexicana and Aeronaves emerged as the two largest carriers, not only in Mexico but in all of Latin America (de Murias, 1989; Taneja, 1988). While CMA was dominant after the war, Aeronaves also grew rapidly by acquiring regional carriers and by gaining prized routes granted by government concession. Meanwhile Mexican state officials forced Pan Am to reduce its holdings of CMA during the war and in 1946 formalized a limitation on foreign equity in national airlines. The move was not surprising. The state had also displayed a nationalist interest in other transportation sectors and air transport was increasingly seen as a strategic industry. Globally a similar trend was taking place and one result was the legitimation of cabotage at the 1944 Chicago Convention (Davies, 1964).

In a move to combat further the market power of CMA and Pan Am, the state also purchased 60 percent of Aeronaves de México. As both majority owner of a national carrier and sole grantor of route concessions, the state would favor the upstart airline over the next 40 years (Davies, 1964; Hanson, 1994). In 1959 the state bought Pan Am's remaining shares to become sole owner of Aeronaves. Later Pan Am would sell its remaining stake in Mexicana in 1967 after a series of financial setbacks. Mexicana began to lose money in the 1950s, culminating in a strike in 1958 that led the government to temporarily take over the company. By 1967 CMA was on the verge of bankruptcy. The state bailed the company out by purchasing ten company-owned airports. Pan Am found a buyer shortly thereafter (Hanson, 1994:202; Jiménez Martínez, 1990:98). Among the private Mexican holders of CMA stock was the Sáenz family (since 1944), which has been very well connected to government and to the big *grupos* (Camp, 1987:181, 201, 204).

At this point the regulatory role played by the state was both activist and growing. In addition to holding sole power over determining domestic routes, the state also decided access to international routes through its negotiation of ASAs. Mexico had, by the late 1960s, established several bilateral agreements including, most importantly, one with the United States (Jiménez Martínez, 1990:96–97). The US–Mexico agreement, dated in 1960 (and officially still in force today, although amended to the point of bearing little resemblance to the original agreement), was a Bermuda-type agreement involving five US and two Mexican carriers. In other words, the ASA, like most between the US and Latin American countries, left most details to the airlines. Unlike other such Bermuda-type agreements, however, routes in the US–Mexico ASA were specified and the bilateral contained a supplementary "screening agreement" that gave each government the power to impede capacity increases by foreign carriers (de Murias, 1989:125). The screening amendment was pushed by the Mexicans in order to protect a dwindling market share held by Mexican carriers, but was later eliminated in 1977. Only 32 percent of international passengers to the country flew on Mexican airlines in 1965 and that figure fell to 26 percent in 1970 (Jiménez Martínez, 1990:224). The fact that Mexican state officials pushed for and got the agreement demonstrates their concern about foreign competition and their commitment to maintaining a national carrier presence.

In sum, at the time the tourism push began just prior to 1970, two domestic carriers dominated the Mexican domestic market. This resulted primarily from the worldwide norms on cabotage combined with the state's willingness to play the roles of regulator and producer. The state controlled one of the two domestic carriers (Aeronaves) and held a small ownership interest in the other. The remaining equity stake was held by domestic private capital. Both national carriers also had significant though dwindling market share of international flights to and from the most important market, the United States. Although several foreign airlines also served the international market to Mexico, they did so as the result of very regimented arrangements (ASAs) and they were prohibited from offering domestic service.

This pattern of development was significantly different from those of hotels and was a product of the capital and technology-intensive nature of the sector as well as its strategic status. Each favored state regulation. In addition, the fact that essentially all governments viewed the air transport sector as strategic in nature made state intervention easier. Government protection of domestic air carriers through restrictive ASAs, subsidies and partial or complete ownership certainly ensured that powerful US carriers would not dominate the market to Mexico. Cabotage ensured that as long as airlines in Mexico were economically viable or economically supported by public moneys, they would remain in operation.

With respect to international air transport the state also demonstrated its willingness to intervene to protect Mexican carriers. This is most evident in the ASA with the United States wherein state officials pushed for and got the screening agreement that could be used to preserve market share. Together cabotage, bilaterals and restrictions on foreign investment all ensured that denationalization of Mexican airlines would not take place. The Mexican air transport sector mirrored the international sector and so did state action. Limiting foreign penetration of airlines was practiced throughout Latin America and for that matter the world. This is not to argue that state actors dutifully reproduced what sector attributes and international norms dictated. Instead these factors strongly shaped incentives for the Mexican public sector but also reinforced and legitimized embedded orientations already present within state agencies.

With the establishment of the tourism poles in Mexico beginning in the 1970s, air travel was recognized by state officials as crucial for meeting goals for incoming tourists. Within Mexico the share of international tourists arriving by air would grow from under 40 percent in 1970 to 60 percent in 1980 (Jiménez Martínez, 1990:161). As the poles became operational, existing international routes added new points and entirely new routes were also added. For instance, US and Mexican negotiators reached agreements on expanded routes to Ixtapa-Zihuatanejo, Cancún, and other coastal cities in a 1973 agreement (FONATUR/INFRATUR nd). Initially, most (and in some cases all) flights to the new poles were offered by Mexican carriers. Scheduled foreign carriers, for example, did not establish routes to Cancún until the resort was well established. The first foreign carriers to the resort, Texas International (later merged with Continental Airlines) and Eastern Airlines, began flights from Houston and New Orleans respectively in 1979. Similarly, as late as 1980 Aeroméxico provided the only international service to Ixtapa-Zihuatanejo (FONATUR nd-d).

One result was an airline "push" domestically, as Aeroméxico and Mexicana more than doubled the size of their fleets. The carriers, which provided the bulk of the traffic to all beach destinations during the 1970s, also upgraded equipment with the purchase of wide-bodied DC-10 and 747 aircraft. This expansion,

Table 6.2: Market Share of International Air Transport to Mexico.[a]

Year	National Carriers	%	International Carriers	%
1970	586	26.0	1,628	74.0
1975	1,265	35.0	2,330	65.0
1979	2,349	39.0	3,695	61.0
1980	2,799	42.8	3,744	57.2
1981	2,858	40.3	4,240	59.7
1982	2,427	42.7	3,256	57.3
1983	3,061	49.3	3,097	50.7
1984	3,586	53.3	3,147	46.7
1985	3,458	54.6	2,871	45.4
1986	3,528	54.6	2,938	45.4
1987	4,372	55.4	3,513	44.6
1988	3,667	46.0	4,307	54.0
1989	3,506	42.0	4,844	58.0

[a]Thousands of Passengers on Scheduled Carriers.
Source: Jiménez Martínez (1990); SCT (1991).

as Table 6.2 shows, aided the domestic carriers in increasing their overall share of the international passenger market from 26 percent to 39 percent between 1970 and 1979 (Jiménez Martínez, 1990:162–163). State-owned Aeroméxico (the name was changed in 1972) increasingly became Mexico's point international carrier, gaining several especially attractive route concessions from the Ministry of Communication and Transport (SCT), the government agency charged with assigning routes. Meanwhile, Mexicana focused primarily on domestic routes and business tourists, and its share of that market grew consistently during the 1970s.

Debt, Deregulation and Privatization

Changes in the Mexican airline sector during the 1980s may be traced to two sets of factors: the debt crisis and the global trend toward airline deregulation and privatization. The former, at least initially, produced contradictory effects. The most immediate result was a sharp decline in air travel to and within Mexico. A severe recession in the country from 1982–1984 sharply reduced international business travel and overall domestic demand. This resulted in financial crisis for the national carriers, Mexicana in particular. Rather than allowing the company to go bankrupt the state took control of the carrier in the mid 1980s by increasing its share of ownership from 14 percent to 54 percent (Hanson, 1994:203). Aeroméxico was also adversely affected by the economic shock, but repeated devaluations of the Mexican peso also produced greater foreign demand for air travel to Mexico as the country suddenly became a "bargain destination" for tourists.

If the initial result of the debt crisis was greater state participation in the domestic airline sector through increased ownership, the longer-term trend was just the opposite. Within five to seven years the state began the process of privatizing the national carriers. Aeroméxico was first to be sold off in 1988 after a short but bitter labor strike culminated in a declaration of bankruptcy. In the process the carrier interrupted service, was reorganized, and then sold to private investors. The bankruptcy, which required state approval, reportedly drew in several high level government officials and involved considerable controversy. The shutdown in service lasted several months, leading SECTUR officials to warn that international arrivals and receipts were in danger (La Botz, 1992:90–104; *The New York Times*, 1988:D8). The strike itself was a result of government efforts to downsize the company. The pending privatization of Mexicana had already been announced and unions feared government attempts to sell aircraft and reduce service were aimed at making Aeroméxico more attractive to potential buyers. The four major unions associated with the airlines were split over government attempts to cut service and jobs. By the time of its reincarnation, the company had shed 90 percent of its 12,000 workers (Sánchez *et al.*, 1993). Privatization of Mexicana followed a year later.

The moves coincided with two larger processes. First was a more general move on behalf of the de la Madrid and Salinas regimes towards privatization. The sell-offs were two of hundreds, including the hotels noted in Chapter 5, as the state reduced its holdings in SOEs from more than 1,100 in 1982 to less than 300 by 1990. Second was a more general policy toward air transport in Mexico, which again emphasized liberalization. In 1987 Mexico and the United States signed a wide-ranging series of amendments to the 1960 bilateral ASA that, among other measures, tripled the number of available routes and increased competition on several existing routes (*Air Transport World*, 1989:40, 44). Mexico initiated a liberalized air policy that it would attempt to integrate into bilaterals in 1990 and by 1991 more regulations were lifted on foreign carriers to Mexico (Hanson, 1994; SECTUR, 1991a). For Mexico, changes in this particular bilateral were especially important since 87 percent of its international market was with the United States (SCT, 1991:65).

The reasons behind privatization and liberalization are multiple and ultimately difficult to separate. In retrospect, however, it appears as though the outcome was overdetermined. Most important were changes in international air transport norms and practices along with a major shift in preferences and resources held by state actors. Globally, privatization and liberalization became a worldwide domino effect stemming from deregulation of the US airline sector a decade earlier (Golich, 1990; Sochor, 1991). While Mexico was an early convert among Latin American nations, the global trend toward opening air markets had already been established. According to aviation analyst Robert Booth, by early 1995 only two major Latin American carriers were still in government hands, and both were slated for privatization (*The New York Times*, 1995:34).

That Mexico was first among many to adopt these measures is best explained at the domestic level. State actors demonstrated a clear commitment to privatization on their own terms as well as reducing government spending. The domestic carriers continued to gain market share on international flights to Mexico, but the government began to argue those gains were too costly. As Table 6.2 demonstrates, the two major Mexican carriers upped their portion of the international airline market to and from Mexico from just under 40 percent to as high as 55 percent during the 1980s. Domestically, the figure ranged between just under 80 percent to more than 90 percent during the same period (Hanson, 1994:203). International flights also provided the main source of income for the national carriers. Mexicana's international service, for instance, made up 38 percent of the carrier's passengers but 55 percent of revenue just prior to privatization (Sánchez *et al.*, 1993:175). Yet for state officials, these figures were less important than budgetary concerns and by 1986 state actors launched a public relations campaign suggesting the national airlines were bloated and losing money. The Budget Ministry (SPP), then under future president Carlos Salinas cut Aeroméxico's budget in 1987 in a first step toward downsizing and later privatizing the company. In contrast, Mexicana's privatization was based on projected future deficits. The carrier was profitable in 1987 and 1988 but was expected to face heavy costs in upgrading equipment in upcoming years (Sánchez *et al.*, 1993:177–178).

Privatization was done in a manner that ensured majority Mexican control over the major domestic air carriers, but also involved some foreign capital. In the case of Aeroméxico, the state sold 65 percent of the carrier to Dictum, a group of Mexican investors, for $300 million. The group also included the pilots' union, which gained between a quarter and a third of the new carrier, as well as a bank and the son of a Mexican president. The pilots had been the lone nonstriking union leading up to the bankruptcy, and publicly supported the declaration of bankruptcy (La Botz, 1992). According to Hanson (1994) and *Air Transport World* (1989), controlling interest in the carrier was sold to Dictum. Bancomer, which had also been involved in hotel investment, was said to have gained 20 percent interest in the airline while a group of investors headed by Icaro Aerotransportes bought the remainder. The buyers were headed by Enrique Rojas and Miguel Alemán. The latter is the son of the famous Mexican president and has ties to grupos Televisa, Novedades, and the hotel chain Posadas (Hanson, 1994; *Latin American Weekly Report*, 1988a).

Mexicana, which had benefited from the Aeroméxico bankruptcy and temporary shutdown, was privatized in two steps a year later in 1989. At the time Mexicana was privatized it controlled 43 percent of the international market to Mexico, and 57 percent of the domestic market and had become the largest airline in Latin America (*Air Transport World*, 1994; Sánchez *et al.*, 1993:175). Continental had 21 percent of the international market to Mexico that year, followed by American (11%) and Delta (5%). These figures are somewhat deceptive in that Mexicana gained significant share in both markets as Aeroméxico first interrupted service and then slowly emerged from bankruptcy. The latter had just five percent of the international market in 1988 and 31 percent of the domestic market. By 1992 those shares increased to 13 percent and 46 percent, respectively (Hanson, 1994:206; Sánchez *et al.*, 1994:176).

The winning bidder for Mexicana, Grupo Falcón, was made up a series of investors headed by the Mexican Grupo Xabre, Chase Manhattan, and European financier Sir James Goldsmith. As part of the winning bid the new owners of Mexicana also purchased a series of other air service companies, including the regional carriers AeroCaribe and AeroCozumel, which gave it greater access to the eastern tourism resort areas of the country (*Air Transport World*, 1989:42).

For Xabre, headed by the Brener brothers, the bid was part of an expansion of tourism services. The Brener family had moved from meat packing and trading in the early 1980s to form Real Turismo, the second largest Mexican hotel group in the country. By the time Xabre gained part of Mexicana, its Real Turismo division owned a series of hotels and the Camino Real and Calinda chains. Real Turismo also owned the Las Hadas resort in Manzanillo at the time it bought Mexicana (*Forbes*, 1989:88). The Breners sold the Camino Real chain in 1993.

Privatization also coincided with deregulation, and this brought new entrants into the Mexican market. US scheduled carriers United, America West, and Alaska Airlines started new routes to Mexico, for example, and a number of regional domestic carriers also appeared (*Air Transport World*, 1989:44; *El Financiero*, 1992e). The full effects of deregulation, have not yet been felt, but as in the US, a contrasting trend has been one of growing concentration. Grupo Falcón consolidated its control when the state completed the second stage of privatization of Mexicana (*El Nacional*, 1992:26; Sánchez *et al.*, 1993:181–182). CMA, which emerged as the weaker of the two domestic carriers and lost between $30 million and $45 million in each of its first three years of private operation, also agreed to a cost-reducing alliance with rival Aeroméxico during that year. The alliance began with sharing of support services, then moved into sharing a CRS system, crews, maintenance, and inventory management (Hanson, 1994:204–205). Aeroméxico also bought eleven percent of the shares in CMA in 1992 and conflicting rumors circulated that one or the other rival would gain control over CMA. The big two carriers, long rumored to become one, finally did engage in a *de facto* merger in early 1993 when a stock swap between them resulted in Aeroméxico controlling 55 percent of Mexicana's voting stock (*Air Transport World*, 1994; Hanson, 1994).

Privatization and liberalization have thus led to conflicting results by the early 1990s. Domestically, concentration has grown as 75–90 percent of the air transport market is controlled by Aeroméxico/Mexicana, in essence one firm that continues to use the names of each airline. The two carriers are not small by any means. According to IATA (1993) figures, in 1992 Aeroméxico was the world's 30th largest carrier in terms of scheduled passengers carried, and Mexicana was 32nd. Considered together they would rank 19th, ahead of carriers such as Qantas, Air Canada, and KLM. On the most busy and lucrative domestic routes the carrier has enjoyed a virtual monopoly (Hanson,

1994:206–207). As Table 6.2 shows, however, their rank has fallen during the decade, to 44th and 50th by 1999.

Nevertheless, the international market as a whole has become more competitive, at least if measured by the number of actors involved in the sector. (Again it is important to note that although many airlines may serve the country as a whole, monopoly or oligopoly conditions may exist within any two city routes.) Amendments to the bilateral with the United States opened new routes and resulted in additional scheduled entrants into the market. More US carriers now serve Mexico, and other foreign carriers, including KLM, Air France, Lufthansa, and Malaysian Airlines, also joined in. By 1990, 31 foreign carriers served Mexico, carrying 4.8 million passengers, or 54 percent of the international market (SCT, 1991:15, 21). One result of this expanded service is that international traffic to Mexico grew rapidly, especially after 1988 (EIU, 1993). Total (domestic and international) passenger arrivals grew 16 percent in 1989, twelve percent in 1990, and 8.5 percent in 1991 (Sectur, 1992:265). Much of this growth went to the tourism poles. Total air passenger arrivals to the five planned resorts nearly doubled between 1988 and 1991, reaching more than 2.4 million. Cancún led the way with 1.7 million and became the third busiest airport in the country in terms of passenger arrivals (SECTUR, 1992:265–266). In terms of international traffic on scheduled airlines, the five poles drew nearly one-quarter of all arrivals to the country in 1991 (SECTUR, 1992:269–270; EIU, 1993:12–13).

Both domestic carriers, already teetering due to consistent annual losses in the early 1990s, were sent into a tailspin by the 1994–1995 peso crisis. The crisis brought domestic air traffic to a virtual halt and Aeroméxico and especially Mexicana found themselves on the verge of bankruptcy. The state again stepped in to ensure their survival when it, along with four domestic banks that were the carriers' major creditors, formed the Corporación Internacional de Aviación or Cintra in 1995. The holding company took over both carriers in June 1996. The banks, which owned 70 percent of Cintra, joined with the state (21%), to jointly manage the two carriers, although their operations have since remained separate in order to conform with Mexican anti-monopoly laws (Knibb, 1996; Levin, 1999).

Aeroméxico and Mexicana faced not only greater competition on international flights during the 1990s, but also the effects of Mexico's economic crisis. One result has been a shrinking of market share. Mexicana and Aeroméxico together held just 39 percent of the international market in 1997 (Knibb, 1996), down from around 42–43 percent earlier in the decade. As Table 6.1 demonstrates, by 1998 Aeroméxico ranked just 44th among the world's busiest airlines as determined by scheduled passenger kilometers flown, and Mexicana was

50th, both down considerably from earlier during the decade. In terms of number of passengers carried Aeroméxico was 27th and Mexicana 37th (IATA, 1999). Both carriers attempted to meet this competition through engaging in strategic alliances with foreign carriers. Aeroméxico and Delta struck up such a codesharing agreement in 1994 that was broadened to include a marketing agreement in 1998 (Delta, 1998). Mexicana reached a similar agreement with United Airlines in 1997 (*Aviation Daily*, 1997).

Conclusion: Changes in the Mexican Airline Sector

While holding some similarities with hotels, airlines have largely followed a distinctive path of development. The airline sector in Mexico has been most affected by early public ownership and regulation, and despite more recent privatization and liberalization it remains less market driven than hotels. Much of this is due to the unique nature of the activity and the remaining international norm of cabotage.

Early state intervention in the airline sector was legitimized both in terms of the strategic interests of the nation and by international norms. The sector originated through foreign investment, but its history is one of heavy state regulation and protection as well as, until recently, significant state participation. The reasons behind state divestiture were many, but despite recent privatization, the airline sector in Mexico remains largely Mexican owned. To be sure, ownership is concentrated among large-scale capitalists including banks and industrial groups, but this is not particularly surprising in such a capital-intensive industry. It is also predictable, given the cozy relationship between the Mexican state and large business interests in Mexico. If anything, the relationship became even more cozy with the wave of privatization beginning after the 1982 debt crisis. Recent privatization of the two major carriers also opened the way for hotel interests to gain a foothold in a related travel business.

In addition to maintaining national control, the state's equity participation and role in regulating domestic and international routes also provided clear advantages for promoting tourism exports. Certainly the state's ownership interest in airlines went back much further and was consistent with ISI and the protection of strategic industry, but it also took on a new role with the tourism export push. Most significant, tourism officials were able to count on air service to the new, untested tourism poles at a time when foreign carriers were absent. Similar to the experience with the publicly-owned hotel chain Nacional Hotelera, the state utilized Aeroméxico and Mexicana to assume initial risks associated with the planned poles while other carriers sat on the sidelines. Even as late as 1988, SECTUR officials warned that the bankruptcy of Aeroméxico could cause severe problems for tourism exports (*The New York Times*, 1988:D8).

Finally, state and subsequent private Mexican ownership mean not only greater control, but also more resources deriving from tourism expenditure. While such figures do not technically include spending on transportation, international pleasure tourism clearly depends heavily upon airlines. The presence of domestic carriers with a fairly high share of the international air market has increased the tourism benefits going to (and staying with) the destination country. In short, a higher proportion of the foreign tourism dollar stays in Mexico.

The other interesting aspect associated with control is that of name recognition, which Chapter 5 argues is so important in the hotel sector. One might expect a similar case for airlines, where for foreigners a familiar name is probably associated with both safety and service. Yet the strict regulatory regime surrounding the granting of routes has, until recently, limited consumer choice for individual routes, thereby making name recognition less important.

The two service subsectors, then, have followed different trajectories over time. This is best explained by industry attributes themselves, a feature discussed more fully in the closing chapter. That chapter also considers the implications for the analytical model and for the larger field of development studies.

Chapter 7

Conclusions: Tourism, Export Promotion and Development

Two goals have guided this book. The first is to present a working model that accounts for outcomes resulting from export promotion; the second is to utilize the model to document the growth and dynamism of one such export industry, tourism in Mexico. Ultimately the goal is for each to inform the other. This closing chapter reconsiders Mexican tourism in light of the analytical model, and then discusses implications for development studies. Finally the book closes with some brief comments on contemporary and future Mexican tourism in the context of NAFTA, Chiapas, the peso crisis, and Mexico's changing political landscape.

Review of Findings

According to Pearce (1989), stage theories have, in the past, been popular among those looking at tourism destinations. Generally the idea pursued is that destinations experience something of a life cycle, starting with discovery, then growth, next maturity, and finally either regeneration, stagnation, or decline. Whether inevitable or even true, stage theories correctly suggest that destinations constantly change, especially in the eyes of potential tourists. In part this is because for many travelers, tourism is a positional good: to some degree enjoyment appears to depend upon excluding others (personal communication with James Mahon; van den Berghe, 1994).

Therefore, it is no surprise that diversification has been the watchword for tourism in Mexico since the 1990s. Today tourism is a dynamic and mature industry in Mexico. It has been at least a decade since the country first demonstrated it could successfully attract tourists in large quantities. For those involved in the Mexican tourism industry, one goal now is retention of tourists: those tourists who may have gone to Ixtapa last year may look for an alternative this year. Will that alternative be another Mexican resort? In addition, new arrivals must be drawn to Ixtapa. Because roughly nine in ten foreign tourists arriving in Mexico continue to originate in the United States, changes in demographics and demand there are also closely watched by tourism providers and officials in Mexico. In this sense promotion never ends, but a quarter century after the tourism push began in Mexico one can take stock of what has changed.

Outcomes are best considered initially at the aggregate level. In 1998 nearly 20 million international tourists traveled to Mexico, spending $7.9 billion. According to recent figures available, Mexico ranks seventh in the world in international tourist arrivals and 11th in earnings (*Travel and Tourism Intelligence*, 1999; WTO, 1999). By 1990 the industry ranked second as employer and earner of foreign exchange, accounting for one in every eleven jobs in Mexico and ten percent of exports excluding oil (SECTUR, 1991a:2). By 1999 one in ten worked in tourism and the industry passed oil and petroleum products in terms of export earnings (EIU, 1999). According to FONATUR figures reported in Madsen Camacho (1996:34), tourism accounts for seven percent of GDP. In short, tourism is today a major industry within the domestic political economy of Mexico. When considered in terms of industrial transformation, which includes the growth noted above but also ownership and control, the picture is less rosy. Although not completely foreign owned or controlled, tourism has become very capital intensive in Mexico and foreign control is significant.

Other commonalities also mark the primary subsectors in question. Hotels and airlines have each undergone liberalization and/or privatization over the past decade. In part this may be traced to the larger process of liberalization undertaken across the Mexican economy since 1982, but it is also the result of a more mature tourism industry where private actors are willing and eager to invest. Despite these, considerable variation remains among subsectors. Hotels retain a complex web of ownership, contractual agreements, and alliances that remain fairly fluid, at least in comparison to the airline sector. There, significant changes have also taken place in the last decade, but air transport remains highly regulated and Mexican firms continue to have a strong presence.

The final outcome worthy of attention is that of the role of the public sector. According to FONATUR figures (1999), the agency invested more than $12 billion in tourism from the period of 1974–1998, split roughly evenly between capital investment and credit offerings. Today, with maturity, the state has largely withdrawn from many of the productive aspects of tourism. While it remains at the forefront in terms of marketing and promotion as well as planning, it no longer owns airlines and has all but moved out of the hotel sector. This leads to two rather paradoxical conclusions: The state, on one hand, has withdrawn from an activist role in tourism compared to the 1970s. On the other, however, it remains more interventionist today in tourism than in many other parts of the economy.

The less interventionist role is, in part, the result of a new dominant developmental ideology, and in this sense tourism mirrors most other areas of the Mexican economy. But the change also must be understood in the context of the industry itself. Much of earlier state action, especially as entrepreneur and banker, was that of surrogate when private investment was lacking. As the industry — and more specifically individual resort sites — matured, private initiative was more forthcoming and the state began to withdraw.

Perhaps the most convincing evidence of the state as surrogate capitalist can be found today in the state's continued selective activism. Despite the market-conforming ideological revolution that has taken place among public sector officials in Mexico in recent years, the state remains comparatively activist in the tourism sector. In 1989 the government of Carlos Salinas launched a 5-year plan calling for modernization and growth of tourism in Mexico. The overarching plan recognized tourism as a strategic sector (Casado, 1997). Aside from continued work on Huatulco, the last of the five planned poles, state activities during the 1990s have centered on launching three ambitious projects aimed at capturing segments of the quickly changing global tourism market. The first two, known as "Colonial Cities" and the "Ruta Maya" (later renamed Mundo Maya) have in effect marked a return to promoting *lo Mexicano*, in that they aim to attract tourists to the interior of the country. Colonial Cities promotes quaint, mid-sized cities in the interior with historical attractions. Mundo Maya, a project involving four other Central American nations and funded in part by the European Union, promotes ancient ruins in southern Mexico (*La Jornada*, 1992c; SECTUR, nd; 1991b). Both projects target history-oriented tourists and promote attractions unique to Mexico and/or Central America. The latter program also explicitly attempts to attract tourists from Europe in an effort toward diversification, although as Pi-Sunyer and Thomas (1997) make clear, contemporary Maya are virtually invisible in the program.

While the first two activities are largely promotional in nature from the standpoint of state actors, the third and most ambitious program has been the ongoing series of Megaprojects. These mini-poles again emphasize diversification in an era of growing competition. At least 15 have been completed by 1998 (Cothran & Cothran, 1998). Some target campers or sport fishing, but the bulk are oriented toward especially affluent tourists. As Chapter 4 summarizes, several contain slips for up to 1,000 yachts, golf courses, and luxury hotels, and condominiums (FONATUR nd-a; SECTUR/FONATUR nd). Many are mini enclaves with little or no connection to surrounding communities. These upscale projects are consistent with the growing trend toward capital-intensive development in tourism documented above. Several sites are adjacent to existing resorts, but others are in undeveloped areas. Projects associated with Mundo Maya and the Megaprojects have come together to aid the development of the so-called Riviera Maya during the 1990s. Today the area, which spreads south from Cancún some 70 miles, contains 10,000 hotel rooms and is linked to inclusive resorts, ecotourist sites and Mayan ruins.

State intervention manifests itself most clearly through reliance on planning rather than leaving growth to the market. In addition, through offering a menu of options to private investors, the state is again willing to take on various roles ranging from promoter, to provider of infrastructure, to financier. Although one important change is that responsibility for developing entire megaproject sites is ceded to investment groups, thereby effectively shutting out all but the largest investors (among the early developers were grupos Sidek, DESC and US developer Donald Koll), the state retains a significant developmental role through offering this menu. FONATUR will, for instance, simply sell the land to a buyer who will carry out the plans for the Megaproject or will finance and build all the infrastructure.

How can this continued state activism in tourism be explained? Much of the answer lies in the nature of tourism itself. First, if stage theories are accurate, and there seems to be at least some truth in them, then tourism sites find themselves continually needing to change in order to maintain their current position. This is particularly true with sun, sand, and sea tourism because Mexico's version is little different than Jamaica's, Cuba's or Aruba's. Reliance on this market requires constant planning and promotional efforts. This, at the national level, requires newness. Similar to Acapulco before it, today a mature Cancún draws a smaller percentage of foreign tourists than before. Those arrivals are also less affluent, with many more arriving via package tour and spending less. Unlike many manufacturing industries, tourism involves consumption *at* and *of* the site itself. In addition, tourists frequently desire to consume sites just once, especially in the sand and sea segment: going to Acapulco once is often enough for many US citizens. Urry (1990) discusses this tendency by considering tourist motivation from the standpoint of semiotics. The fundamental motivation and activity of individual tourists is to collect and consume signs, attached to which are meanings. Once these signs are collected, however, there may be little reason to return to a given site. Consistent with this, but at a much more practical level, several public sector officials interviewed indicated that one of the biggest problems facing the domestic industry was getting repeat business.

At the national level, then, state actors have an interest in fostering renewal. They have little left to do in places like Cancún, where based on their criteria, success has been achieved. Such places may be left to the private sector. Yet the need for renewal demands new state efforts. Therefore, they become involved in further promotion and planning, and because renewal frequently involves new risks, the state again finds itself acting as banker or surrogate entrepreneur. This is not to suggest that state actors *must* do this; state managers in Mexico could withdraw tomorrow. Sun and sea based tourism, however, invites more state intervention, and thus state actors are on a treadmill of sorts, running in place to maintain Mexico's reputation for new, world-class sun and sea resorts. Meanwhile, what happens to the "old" resorts also becomes a question state officials are not particularly eager to address. If Acapulco is any indication, international resorts slowly become less attractive "national" resorts and entropy. Natural resources and historical sites are consumed and then discarded, and as they become more scarce they become controlled and commodified as tourism sites.

Explaining Industrial Transformation

The last four chapters have not attempted to *prove* the underlying truth of the model, but instead to link facts with analysis. The true test is how well the two fit together and whether an alternative account is more convincing. The outcomes summarized above did not happen by accident. Instead they are traceable primarily to the factors laid out in Chapter 2, especially the activist role of the state in violating market signals when it came to tourism, and industry characteristics that shaped distributional patterns of ownership and control. This claim suggests that a generalizable logic exists that accounts for the changes that have taken place in Mexican tourism.

State-led Growth

There is little doubt that tourism constituted an export push and as such the state was a central actor in shaping the industry. The particular model of tourism development taking place in Mexico is a product of a "tourism imagination" originating in the minds of central bank bureaucrats. This particular imagination linked large scale development of grand, international class resorts with developmentalism and modernity. The resulting pattern during the last quarter century reflected a vision held by state actors housed in FONATUR and

SECTUR. That vision was characterized by planning, by large-scale developments, and by export-orientation. In addition to putting forth their own vision of what tourism in Mexico should be, state actors also worked in a manner that fostered their own imagined view of what development constituted. At the most basic level they promoted growth: public sector funds poured into the industry, especially in the early and mid-1970s, in order to provide the necessary infrastructure for such projects. The state also took the lead in marketing tourism abroad through advertising campaigns and establishing promotional offices.

Furthermore, state actors played crucial roles in production and finance. Directly, public enterprises such as the hotel chain Nacional Hotelera became a major factor in expanding the supply of hotel rooms, and more important, in offering lodging in the untested planned resorts. Government-owned airlines played a similar if less dramatic role. As banker, the state aided in making tourism more attractive to the private sector by setting up and paying for preferential financing arrangements. As Chapter 4 demonstrates, this was especially the case in the hotel sector where public funds built the bulk of hotel rooms over the past 20–30 years. As a result, by the early 1990s Mexico had more than twice as many hotel rooms as any other country in Latin America and ranked seventh globally in lodging capacity (Sussman & Cooper, 1993; WTO, 1994:25).

The activist role played by the state can be tied to two factors: capacity and autonomy. Internally, state actors held sufficient power, expertise, and access to material resources to be effective. Internal conflicts did arise, notably between officials at FONATUR and SECTUR, but there is little evidence that the overlapping responsibilities among the bureaucracies were detrimental. In terms of autonomy, domestic private sector interests surrounding tourism remained, in the 1960s and 1970s, relatively unorganized. State promotion of the planned beach destinations, for instance, certainly threatened to divert tourists from existing resorts such as Acapulco, and yet efforts to combat or alter the planned poles were noticeably absent. In short, the state had more room to maneuver before tourism became a mature industry.

The lack of societal organization is due primarily to two factors prevalent at that point. First, tourism was still small and disparate economically. Owners of hotels or other tourism-related businesses were both relatively few in number, and geographically dispersed. Many were located in Mexico City, a destination that was not directly threatened by plans for the poles. Second, providers of tourism services consisted of primarily small and medium-sized businesses. Outside the airline sector, which is much more capital-intensive in nature, the large-scale *grupos* were still largely on the sidelines. Opposition was nonexistent or confined to small, geographically isolated groups surrounding the poles themselves. These factors, combined with the material resources made available by loans from multilateral institutions, allowed state officials largely to put their own project into place without outside interference.

In sum, although domestic and international business groups clearly benefited from tourism development in Mexico, it would be an oversimplification to suggest that state action was motivated solely by those interests. The state may have ultimately served the interests of capital though promoting tourism, but public sector actors had their own motivations for pursuing such plans. In part they faced genuine macroeconomic — and political — constraints that demanded action. In addition, the particular developmentalist imagination held by key actors, an imagination that linked large scale projects, state control and the exploitation of natural resources in the name of modernity, also served as a basis for such plans. Moreover, state actors themselves benefited from tourism development, both politically and monetarily. In short, the state was not solely motivated by a need to do the bidding of capital.

That the state was activist here runs counter to expectations associated with neoclassical approaches to political economy. One such expectation is that cases of industrial dynamism tend to be the product of market forces with the state taking on a minimalist role. Here, however, the state took great pains to distort market signals. Second, and related, new political economy variants contend that state intervention in the economy commonly leads to suboptimal outcomes or "government failure" (Buchanan, Tollison & Tullock, 1980; Krueger, 1990). However, by most neoclassical measures — and measures adopted by state officials themselves — the industry has been a success. Tourism has demonstrated steady growth, is largely in the hands of private actors, and is subject to few trade barriers. Again, proponents of a minimalist state are thus left only with an unlikely counterfactual: reliance on market forces and no government intervention would have produced a more impressive record.

Such a scenario is difficult to imagine. Undoubtedly tourism in Mexico would have grown without the state leading the way. One cannot ignore or downplay the overall growth in global tourism, which certainly would have included Mexico, and related geographical considerations. Mexico's proximity to the United States all but insured growth in tourism. One can only speculate about what would have been, but it is difficult to imagine tourism growing as fast as it has with minimal state involvement. In addition, market-driven tourism most likely would be more concentrated near the US–Mexican border. It also would probably have resembled the more polluted and overcrowded Acapulco than Ixtapa or Huatulco. Furthermore, because of industry-specific advantages, distributional patterns may have become even more concentrated and with less Mexican participation.

That the Mexican state was interventionist is important for the analysis here, but equally important is answering why it was more "developmentalist" (or capitalist) in this particular case rather than predatory. This is especially perplexing because simultaneous state intervention elsewhere in the economy has been viewed, *post hoc*, as inefficient and in many cases corrupt. In Evans' (1992) words, why was "the state" the solution in the tourism industry at the same time that it was the problem elsewhere? The answer is complex and multifaceted, and to a certain extent components of that answer are best considered cumulative in nature.

First it should be said that these oppositional attributes cannot by themselves account for outcomes. As ideal types they are useful but they also may dangerously evolve into caricature. Undoubtedly the Mexican state was not purely and at all times developmentalist when it came to tourism. As the concluding part of Chapter 4 indicates, officials did make mistakes and at times even became involved in corrupt practices. Yet as Evans (1995) points out, the line between autonomy and embedded autonomy is a fine one. He suggests that because autonomy alone may easily lead to predation, some links between state officials and societal groups are necessary for "successful" developmental outcomes. In addition, corruption is not limited to activist states. One interesting aspect in the Mexican case is what appears to be the rising incidence of corruption and predation under the privatizing state during the 1990s. Ex-President Carlos Salinas routinely proclaims his innocence these days from his adopted home in Ireland, a country with which Mexico has no extradition treaty. Some of this graft — involving senior government officials — has reportedly involved real estate in tourism zones and possibly some air transport.

Despite these caveats several factors aided the state's more positive contribution to tourism development. One was the very nature of the project. The macroeconomic stimuli outlined in Chapter 3 encouraged exports, which in itself invited greater efficiency. International developments also made tourism something of an obvious choice. By the 1960s global tourism was booming, and Mexicans benefited from being adjacent to the largest tourism market in the world. Moreover, government officials in industrialized countries along with multilateral institutions had by this time jumped on the tourism bandwagon, arguing it was a means for economic development.

As a result, multilateral institutions made tourism a more attractive choice for Mexican policymakers through their offers of preferential financing. The availability of concessional loans, made tourism an easier choice for them. Mexico was at the forefront of this, gaining some of the first multilateral loans explicitly for tourism development. At a more general level, however, state officials did not attempt something completely novel when they embarked on tourism export promotion. Several other countries, many in the Third World, also encouraged tourism at this point. What was novel, however, was the scale involved. State actors set ambitious goals, such as attracting a million tourists a year to one site alone (Cancún), and making Mexico one of the most popular destinations in the world.

Despite these influences, however, state actors still could have been predatory. Richter (1989:51–52), for instance, found similar features at work in the case of the Philippines, where the outcome of tourism export promotion was corruption, waste, and cronyism. This industry there ultimately became a political instrument for President Ferdinand Marcos that rewarded supporters, enriched his family, and served as symbolic support for his regime. Additional factors in the Mexican case — this time at the domestic political level — also aid in explaining the developmental aspects of state action. For example, institutional arrangements and staffing of state tourism agencies fostered a more positive role. INFRATUR and later FONATUR were insulated from both societal and other institutional interests by their early location within the central bank. Moreover, staffing the agencies mainly with bank officials provided a sense of cohesiveness and shared purpose. Foremost among staff goals or embedded orientations was a commitment to increasing export earnings, a priority that shaped the type of development pursued for years to come. In this sense it is perhaps more useful to shift levels of analysis from developmental states to bureaucracies.

The Limits of Statist Explanation

While tourism clearly resulted from aggressive state action, one of the main points of emphasis that emerges from the analysis here is that attention to the state is not enough for explaining the outcomes of industrial transformation. It is also crucial to point out that state action was structurally limited at the outset, and became even more limited with time. Because officials pursued foreign tourists on a large scale (in other words, mass tourism) certain options were in effect closed off. Here factors associated with international industrial organization became especially important, and they also varied among subsectors.

As Chapter 5 suggests, for example, drawing in large numbers of foreigners meant the state first needed to attract foreign hotel chains. Name recognition, trust, and ties to the tourist generating market sent clear signals. Because the nature of the product itself involved high risk for consumers, and because those consumers could easily substitute Jamaican beaches for those in Mexico, the chains were crucial for the new tourism poles. Therefore, state action was most constrained in hotels, where name recognition held the strongest appeal to foreign tourists.

The outcomes in development patterns at a subsectoral level were somewhat overdetermined. In large part they are traceable to industry characteristics and the reproduction of international structures on a domestic level. As globalization, integration, and centralization have affected the international tourism business, this tendency to reproduce the global locally has become even more likely. The simple requirements of scale have increased, as have the needs to link businesses to global information and marketing networks. The state aided in this outcome through emphasizing mass tourism back in the 1960s. While this did not make the developmental patterns three decades down the road inevitable, in hindsight it appears to have made them likely.

Models of Industrial Change and Development

The most basic finding of this study — that the tourism industry in Mexico is much larger and dynamic, as well as more organized and "capitalist" than 35 years ago — is not particularly surprising. More contentious may be the alleged forces behind this transformation. The state was activist but did it "cause" the outcomes summarized above? Were industry characteristics as constraining as they are made out to be?

Many of the conclusions reached here are structural in nature, thereby making them susceptible to what Haggard (1989, 1990) refers to as the structuralist fallacy. Here structures may be broadly asserted and by their very assertion said to constrain action. The problem is they are seldom measurable and thus difficult to falsify. Even if one accepts that structures are constraining, this situation is open to interpretation (Adler, 1987; Haggard, 1989).

One such conclusion drawn in the present study is that international market structures associated with tourism shaped local development patterns in the respective subsectors. In addition, they constrained state action. As a result, for example, one central argument is that the international tourist-class hotel subsector became dominated by a multinational class alliance more or less regardless of state preferences. At least three other possibilities exist: the outcome could have reflected state preferences and resulted from conscious state action, it may have happened as an accidental consequence of state action, or it could have had nothing to do with either international market structures or state action. The problem is the lack of a "smoking gun". In other words, there are few, if any, observable indications of these possibilities that would settle the issue once and for all.

The structuralist fallacy is a very real danger, but so is erring on the opposite side of the spectrum. Difficulty in observation of social phenomena is not a new problem and it does not mean that these phenomena fail to exist. Instead three things can aid in drawing such structural conclusions. First is attention to context. In the present case, for instance, one of the most striking aspects is that the same state that was simultaneously acting to broadly limit foreign investment elsewhere in the economy welcomed it in tourism, and especially in hotels. The logical question is why. The most compelling explanation is that the costs of bypassing foreign capital were simply too high, especially in the accommodation sector where name recognition for foreign tourists appears to have been most important. Welcoming transnational chains does not prove beyond a doubt, but does strongly suggest that for some reason Mexican state officials desired a foreign hotel presence. Second is to consider counterfactuals, which Gereffi (1983) does in a compelling manner with his study of the pharmaceutical industry and development in Mexico. In the tourism case, one may ask what would the outcome have been if the state had maintained the exclusion zones and bypassed foreign participation? Certainly the answer is open to speculation, but again it is difficult to imagine nearly 20 million foreign tourists traveling to Mexico each year without well-established links to the international industry.

A third factor affecting whether structural arguments are convincing is the nature of the study itself. One of the best strategies is to draw conclusions from a large-N study. Such measures control for anomalies and thus provide the researcher with greater confidence when drawing conclusions. Often, however, this is difficult. Certainly a trade-off is involved between the number of cases included and the depth in which one may examine them. This book represents both a case study and a small comparative study. It is a case with respect to the first question asked — the source of export dynamism — in that it looks at one economic activity within one country. It constitutes a comparative study, however, in regard to the second question. In answering what industrial transformation looks like and why, it looks at both hotels and airlines in order to highlight how different industrial structures influence local development.

Certainly greater confidence could be asserted if the findings here were to be replicated or broadened. This may be done both by studying tourism elsewhere and also examining other export industries. While calling for further study is predictable to the point of being trite, there is also a sense of urgency involved. As is noted at the outset of this study most countries in the world are embracing export-led growth strategies. While much of the focus is on manufacturing, it is by no means the only emphasis. From a development policy standpoint, one requires a much deeper and nuanced understanding of what can be expected from such strategies.

At least two policy implications derive from this study. First the findings here call into question policies associated with the so-called neoliberal ascendency by arguing that state intervention was necessary for achieving the tourism growth recorded. This suggests that state activism is warranted in economic activities where significant bottlenecks exist *and* when state institutional arrangements are sufficient for carrying out necessary roles. The second, and somewhat paradoxical conclusion is that even under such conditions, the developmental prospects for this model of tourism are still significantly limited in terms of distributional gains. One might argue that emphasis on ownership and control are rather narrow and that job creation, export earnings, and other spillover effects contribute substantial developmental benefits. Here, however, tourism may offer less than other economic activities. Frequently the industry can take on an enclave nature, establishing few linkages with the larger economy. Moreover the majority of jobs are seasonal and comparatively low paying. For these reasons ownership is especially important for maximizing benefits.

The implication is not that tourism exports should never be promoted. Frequently few alternatives exist. On the other hand, the type of tourism promoted may be challenged as being capital intensive and offering fewer local benefits than was once believed. The massive resources poured into Mexico's experiment with large-scale development trickled down to few ordinary Mexicans. It is not difficult to think of alternative uses for these resources.

While less industrial tourism may be warranted, however, mass tourism continues to appeal in many countries. Even in Mexico, a relatively "wealthy" Third World country with a diversified economy, the industry offers employment opportunities in a nation where roughly one million additional people enter the workforce each year. For students of development the task is not to label policies as good or bad in retrospect. Instead, the point is to highlight the range of developmental possibilities and constraints.

At another level, this study crosses two subfields of political science that appear to be moving in opposite directions while engaging in little dialogue. Within comparative politics, studies of political economy have embraced the state as the primary causal element for all kinds of developmental outcomes. In a sense this trend has turned 180 degrees from the heyday of dependency theory that emphasized the international constraints on development. Much of the literature on NICs and would-be NICs, for instance, seems to suggest that states tend to get what they deserve. Good policies beget good outcomes and bad ones the opposite. At the same time a considerable body of political economy literature within the international relations subfield claims that due to globalization the state is losing both political power and analytical weight when it comes to development. There are twin dangers in going too far in either direction and in ignoring the opportunity for dialogue.

The findings of this study are by no means confined to tourism or export pushes. Instead the model and the study amount to a call for greater integration between international and domestic determinants of development. While the "return to the state" rightly forced analysts to stop ignoring domestic politics, this does not mean powerful political and economic structures suddenly ceased to exist. The opposite danger, however, is to label globalization as a homogeneous force that constrains political action in some uniform manner. Instead what is needed is to specify the possibilities for state action in a given policy area, as well as the forces that limit such action.

Studying individual economic activities or industries is particularly useful for drawing mid-level theoretical conclusions having to do with development. Most policy does not take the form of something one later calls development strategies and is not formulated by a unitary actor. Instead it is most often aimed at particular industries, and done by individuals housed in agencies that only later are regarded as "the state". That is why in this case it is perhaps more accurate to draw conclusions about a developmentalist bureaucracy rather than a developmentalist state. Similarly, the process of internationalization varies by sectors and industries, not just temporally but also structurally. Drawing overarching conclusions about either states or globalized markets must be avoided. Instead a grounded examination of state attributes and choices combined with industrial structures offers a more balanced and ultimately accurate picture of the forces shaping development.

Conclusion: The Future of Tourism in Mexico

Much has changed recently, both politically and economically, in Mexico. How these changes might affect the future of tourism is worthy of discussion. The growth of this industry will most likely continue at a rapid pace in Mexico, although broader economic and political contingencies now and in the future certainly will have a profound effect on foreign tourism. Four such factors have arisen since 1990: the North American Free Trade Agreement (NAFTA), the Chiapas uprising and related questions of political stability, the 1994–1995 peso crisis, and rising crime and issues of tourist safety. All of these and more are worthy of discussion. NAFTA, which went into effect on January 1, 1994, somewhat surprisingly (and unlike the earlier Canada–US Free Trade Agreement) does not specifically address tourism. The agreement will have long-term effects on the industry most directly in transportation, and more specifically in the previously heavily regulated bus sector which is being opened to foreign competition. Air transport was also left out of the agreement and will continue to be dealt with through bilateral ASAs. The most significant impact from NAFTA on tourism will likely be indirect, in the form of a greater business market. In the 1990s the most competitive aspect of the hotel sector has been in mid-sized cities where new hotels are popping up in hopes of tapping into this market.

Perhaps more significant than NAFTA in the short to medium term are the peso crisis and the uprising in Chiapas. The former crisis, beginning in December 1994, has resulted in the value of the Mexican currency falling from roughly 28 US cents to around ten. For foreign tourists, economic crisis often translates into bargains and this took place to some degree in Mexico shortly after the crisis. It is important to note, however, that many traditional resorts are "dollar destinations", where almost all vacation components are priced in US dollars. The more direct and immediate effect was on tourism investors. Grupo Situr, like many Mexican groups, held considerable foreign denominated debt and defaulted on repayment obligations (*The New York Times*, 1995). The Camino Real hotel chain, one of the largest in the country, went bankrupt and was auctioned by the state in 2000. Other hotel firms halted all new construction in the aftermath of the crisis (author interview, 1995) and overall investment in tourism slowed considerably until 1998. Banks found themselves owning properties, especially hotels, after a series of loan defaults due to the crisis. Subsequently several banks found themselves insolvent and were bailed out by the state. New Mexican government agencies had been charged with selling off many of the assets (in a manner similar to that of the savings and loan crisis in the United States) that had been acquired by the state as a result of this process, including several properties. Today much of the new hotel investment in Mexico is not for new construction, but rather the sale of these properties to private investors (*Hotels*, 1999).

A more serious potential threat to tourism in Mexico is political instability. The uprising in the southern state of Chiapas, combined with political violence elsewhere in the country could, if it reignites, spell disaster for the industry. In 1996 another guerrilla group made brief but violent strikes in Guerrero state, not far from Acapulco, as well as the resort of Huatulco. As experiences in Egypt, Yugoslavia, Lebanon, Fiji, and several other countries have clearly demonstrated, foreign tourists demand security and safety. Even the slightest hint of danger could destroy a destination. Even without violence, the status of the long-ruling PRI also raises questions of stability. Having controlled Mexican politics for 70 years, the party began losing popular support in the 1980s and 1990s, including a series of local and regional elections. Finally, in

July 2000, Vincente Fox, of the conservative opposition National Action Party (PAN) won a historic presidential election.

In Mexico's strong presidentialist political system, the impact of the PRI losing control of the executive branch cannot be overstated. Some suggest that the party will likely die off in the future, although it has demonstrated amazing resilience in the past. With respect to the impact on tourism in Mexico, however, many questions remain. Uncertainty is the enemy of both tourists and investors and as a result any change of regime could lead to at least short term negative consequences in this industry. At the same time it is nearly impossible to distinguish the PAN's pro-market economic platform from that of the PRI in recent years. Certainly the question of stability and peace during this transition period will have the most profound potential impact on the future of tourism in Mexico (Cothran & Cothran, 1998).

Perhaps the biggest threat to tourism in Mexico today is real and perceived threats to the personal safety of foreigners. As drug trafficking has taken a serious foothold in the country over the past decade, several areas of the country have become much more violent. The border regions in particular have become much more dangerous and, in addition, drug gangs appear to have infiltrated both police and political offices. Drug money is also widely rumored to have taken over some tourism properties, especially in Cancún, Acapulco, Puerto Vallarta, and Ixtapa. Further, white collar corruption has directly affected the industry. In 1994 the chairman of Aeroméxico became a fugitive and was charged with embezzling more than $70 million. He claimed he had contributed it to the electoral campaigns of the PRI. Another fugitive, this time a banker, took control of the Camino Real hotel chain in the early 1990s. Later, a leading tourism official, Sigfrido Paz Paredes, was charged with illegally receiving profits from state tourism projects in 1997. In Spring 2000, the former governor of the state of Quintana Roo (where Cancún is located) was indicted and became a fugitive from justice.

In addition to this criminal activity, street crime in Mexico City increased exponentially after the peso crisis. Foreigners and wealthy-looking individuals increasingly became targets of kidnappers and robbers. The US State Department responded by issuing travel warnings to its citizens as early as 1997. In 1998 an American businessman was kidnapped and killed in a high-profile incident. As a result, the capital has moved from being one of the safest cities in the world just a decade ago, to one of the most dangerous (EIU, 1999). Although the traditional beach tourism zones have yet to feel the same threats to safety prevalent in the capital or some border regions, the mere perception of Mexico as a dangerous destination constitutes a serious threat to overall tourism growth.

Beyond these stark real and potential changes, a quieter transition is also taking place within the industry. As the national market matures, it appears to be splitting along two lines. Diversification targets the increasingly segmented market through the Megaproject, Colonial Cities, and Mundo Maya programs referred to above. Mexico is also on a golf course building boom since the second half of the 1990s and talk of allowing casino gambling crops up regularly. The second line is aimed at the more traditional mass tourists seeking a beach destination. The maturing of this segment of the market has led to greater competition — both domestically and internationally — and subsequently has contributed to a more significant presence by tour operators who combine price and convenience through package tours. According to one observer, the Riviera Maya stemming down from Cancún has become "the kingdom of all-inclusive hotels and the wholesale tour operators" (Quoted in *Hotels*, 1999:48).

Inclusive tours (ITs) or packages have been controversial among tourism analysts for some time. For some they homogenize a destination and represent a first step toward overrunning and even ruining it. For example, EIU (1993:22) states that with the price of ITs coming down, Mexico is "beginning to attract the wrong kind of tourist or, at least, too many budget tourists". For defenders, this is elitist. Instead ITs, if anything, democratize a destination by making it more accessible to a wider segment of society. In fact ITs likely do both. While making destinations more affordable, they also encourage what might be called the "enclavization" of tourism. By providing a whole range of services and spaces for tourists interested mainly in beaches and sun, they leave guests with little need to leave the resort. It is quite evident that ITs are flourishing in mature destinations. Thus, it is not surprising, in this sense, that Acapulco and Cancún are the two most popular IT destinations in Mexico.

As the Mexican market matures, inclusive tours will most likely continue to grow, as will other price-competitive alternatives to traditional tourism arrangements. For consumers, lower prices and the convenience hold great appeal. For providers of services, however, two sets of implications are troubling. First, as price competition intensifies, the material benefits derived from tourism exports are likely to dwindle. Increased volume may not make up for slashed profit margins. Second, those who design and control packages are based in the sending market. What little control local hoteliers and other providers of services held is likely to decrease.

In the final analysis, unless it is derailed by political or economic instability, the Mexican tourism industry will likely continue to grow at a rate roughly equal to global averages. As a result it will probably outpace aggregate growth and employment creation. In sum, this industry will continue to be a central element of the Mexican political economy. Beneath its steady growth, one may expect a continued deeper transformation, however, one of increased integration into global tourism. In terms of who benefits from this transformation, the Mexican industry does not appear to be threatened by denationalization. Domestic firms continue to hold a strong ownership stake within many areas of this multi-sector industry. Instead, if the past is any indication, the organization of the domestic industry — and the bulk of the benefits derived from it — will be marked less by nationality and more by social class.

References

Acuña, Jáuregui, and Mónica de la Garza
  1989   La Dimensión Territorial del Proyecto Cancún, Estado de Quintana Roo. *In*
  Teoría y Praxis del Espacio Turístico, D. Hiernaux, ed., pp. 121–133. Mexico City:
  UAM-Xochimilco.
Adler, Emanuel
  1987   The Power of Ideology: The Quest for Technological Autonomy in Argentina
  and Brazil. Berkeley: University of California Press.
Air Transport World
  1989   Air Transport World 26(10):40–45.
  1994   Air Transport World 31(6):160–165.
Alarcón, Diana, and Terry McKinley
  1992   Beyond Import Substitution: The Restructuring Projects of Brazil and Mexico.
  Latin American Perspectives 19(2):72–87.
Alisau, Patricia
  1992   Grupo Posadas Corners the Hotel Market. Business Mexico 2(6):12–14.
Ames, Barry
  1987   Political Survival: Politicians and Public Policy in Latin America. Berkeley:
  University of California Press.
Amsden, Alice
  1979   Taiwan's Economic History: A Case of Etatisme and a Challenge to Dependency Theory. Modern China 5(3):341–380.
  1989   Asia's Next Giant: South Korea and Late Industrialization. New York: Oxford
  University Press.
  1992   A Theory of Government Intervention in Late Industrialization. *In* State and
  Market in Development: Rivalry or Synergy? L. Putterman and D. Rueschemeyer,
  eds, pp. 53–84. Boulder CO: Lynne Rienner.
Ascher, François
  1985   Tourism: Transnational Corporations and Cultural Identities. Paris: UNESCO.
Aviation Daily
  1997   Aviation Daily 329(July 29):173.
Aviation Week and Space Technology
  1994   Aviation Week and Space Technology 141(October 5): 131.
Baer, Werner
  1972   Import Substitution and Industrialization in Latin America: Experiences and
  Interpretations. Latin American Research Review 7:95–122.
Bailey, John J.
  1988   Governing Mexico: The Statecraft of Crisis Management. New York:
  St. Martin's Press.

Bain, Joseph
  1968 Industrial Organization. New York: Wiley.
Balassa, Bela, Gerardo M. Bueno, Pedro-Pablo Kuczynski and Mario Henrique Simonsen.
  1986 Toward Renewed Economic Growth in Latin America. Washington DC: Institute for International Economics.
Banco de México
  1942 El Turismo Norteamericano en México, 1934–1942. Mexico City: Banco de México.
  1993 Annual Report, 1992. Mexico City: Banco de México.
Baretje, René
  1982 Tourism's External Account and the Balance of Payments. Annals of Tourism Research 8: 57–67.
Barham, Bradford, Mary Clark, Elizabeth Katz, and Rachel Schurman
  1992 Nontraditional Agricultural Exports in Latin America. Latin American Research Review 27(2):43–82.
Barnett, Michael N.
  1992 Confronting the Costs of War: Military Power, State, and Society in Egypt and Israel. Princeton NJ: Princeton University Press.
Bates, Robert H.
  1981 Markets and States in Tropical Africa: The Political Basis of Agricultural Policies. Berkeley: University of California Press.
Bennett, Douglas C., and Kenneth F. Sharpe
  1982 The State as Banker and Entrepreneur: The Last Resort Character of the Mexican State's Economic Intervention, 1917–1970. *In* Brazil and Mexico: Patterns in Late Development, S. A. Hewlett and R. S. Weinert, eds, pp. 169–211. Philadelphia: Institute for the Study of Human Issues.
  1985 Transnational Corporations Versus the State: The Political Economy of the Mexican Automobile Industry. Princeton NJ: Princeton University Press.
Bennett, M., and M. Radburn
  1991 Information Technology in Tourism: The Impact on the Industry and Supply of Holidays. *In* The Tourism Industry: An International Analysis, M. T. Sinclair and M. J. Stabler, eds, pp. 45–65. Wallingford: CAB International.
Bhagwati, Jagdish
  1978 Anatomy and Consequences of Exchange Control Regimes. Cambridge MA: MIT Press.
  1987 International Trade in Services and its Relevance for Economic Development. *In* The Emerging Service Economy, O. Giarini, ed., pp. 279–289. Oxford: Pergamon.
Biersteker, Thomas J.
  1987 Multinationals, the State, and Control of the Nigerian Economy. Princeton: Princeton University Press.
  1995 The "Triumph" of Liberal Economic Ideas in the Developing World. *In* Global Change, Regional Response, B. Stallings, ed., pp. 179–196. New York: Oxford University Press.
Block, Fred
  1977 The Ruling Class Does Not Rule: Notes on the Marxist Theory of the State. Socialist Revolution 7(May–June):6–28.

1987 Revising State Theory: Essays in Politics and Postindustrialism. Philadelphia: Temple University Press.

Boniface, Brian, and Chris Cooper
1994 The Geography of Travel and Tourism (2nd ed.). Oxford: Butterworth Heinemann.

Bosselman, Fred P.
1978 In the Wake of the Tourist. Washington DC: The Conservation Foundation.

Bradford, Colin I., Jr.
1990 Policy Interventions and Markets: Development Strategy Typologies and Policy Options. *In* Manufacturing Miracles: Paths of Industrialization in Latin America and East Asia, G. Gereffi and D. L. Wyman, eds, pp. 32–51. Princeton NJ: Princeton University Press.

Britton, Stephen G.
1981 Tourism, Dependency and Development: A Mode of Analysis. Development Studies Center Occasional Paper No. 23. Canberra: The Australian National University.
1982 The Political Economy of Tourism in the Third World. Annals of Tourism Research 9:331–359.
1991 Tourism, Capital and Place: Towards a Critical Geography of Tourism. Environment and Planning D: Society and Space 9:451–478.

Buchanan, James M., R. D. Tollison, and Gordon Tullock
1980 Toward a Theory of the Rent-Seeking Society. College Station: Texas A&M University Press.

Bull, Adrian
1991 The Economics of Travel and Tourism. Melbourne: Pitman.

Burden, William A. M.
1943 The Struggle for Airlines in Latin America. New York: Council on Foreign Relations Press.

Business Mexico
1988 Business Mexico 5(1):66–69.

Caballero, Gloria
1984 Historia Legislativa del Turismo en México. Mexico City: Instituto Mexicano de Investigaciones Turísticas.

Camp, Roderic A.
1989 Entrepreneurs and Politics in Twentieth Century Mexico. New York: Oxford University Press.

Cardoso, Fernando Henrique, and Enzo Faletto
1979 Dependency and Development in Latin America. Berkeley: University of California Press.

Carnoy, Martin
1984 The State and Political Theory. Princeton NJ: Princeton University Press.

Casado, Matt A.
1997 Mexico's 1989–1994 Tourism Plan: Implications for Internal Political and Economic Stability. Journal of Travel Research 36(1):44–51.

Casar, Jose
1988 An Evaluation of Mexico's Policy on Foreign Direct Investment. *In* Mexico and the United States: Managing the Relationship, R. Roett, ed., pp. 37–50. Boulder: Westview Press.

Cason, Jeffrey
  1993   Development Strategy in Brazil: The Political Economy of Industrial Export Promotion, 1964–1990. Ph.D. dissertation, University of Wisconsin-Madison.
  1994   Is there at Tiger in Latin America? Brazilian Export Promotion in Comparative Perspective. Paper presented at the 18th International Congress of the Latin American Studies Association, Atlanta.
Caves, Richard E.
  1974   International Organization. *In* Economic Analysis and the Multinational Enterprise, J. H. Dunning, ed., pp. 115–146. New York: Praeger.
Centeno, Miguel Angel
  1994   Democracy Within Reason: Technocratic Revolution in Mexico. University Park: Pennsylvania State Univeristy Press.
Chant, Sylvia
  1992   Tourism in Latin America: Perspectives from Mexico and Costa Rica. *In* Tourism and the Less Developed Countries, D. Harrison, ed., pp. 85–101. London: Belhaven.
Clancy, Michael
  1998   Commodity Chains, Services and Development: Theory and Preliminary Evidence from the Tourism Industry. Review of International Political Economy 5(1):122–148.
Clarke, Simon, ed.
  1991   The State Debate. New York: St. Martin's Press.
Cockcroft, James D.
  1983   Mexico: Class Formation, Capital Accumulation, and the State. New York: Monthly Review Press.
Cothran, Dan A., and Cheryl Cole Cothran
  1998   Promise or Political Risk for Mexican Tourism. Annals of Tourism Research 25(2):177–197.
Cowan, Ruth Anita
  1987   Tourism Development in a Mexican Coastal Community. Ph.D. dissertation in anthropology, Southern Methodist University.
Cumings, Bruce
  1987   The Origins and Developments of the Northeast Asian Political Economy. *In* The Political Economy of the New Asian Industrialism, F. Deyo, ed., pp. 44–83. Ithaca NY: Cornell University Press.
Cypher, James M.
  1990   State and Capital in Mexico: Development Policy Since 1940. Boulder CO: Westview.
Daltabuit, Magalí, and Oriol Pi-Sunyer
  1990   Tourism Development in Quintana Roo, Mexico. Cultural Survival Quarterly 14(1):9–14.
Davies, R. E. G.
  1964   A History of the World's Airlines. London: Oxford University Press.
Davis, Diane E.
  1994   Urban Leviathan: Mexico City in the Twentieth Century. Philadelphia PA; Temple University Press.
Delta Air Lines
  1998   Delta and Aeromexico Plan Expanded Relationship, Including a Five-Year Marketing Agreement. Press Release, March 10.

de Mateo, Fernando
1987 México: Una Economia de Servicios. Informe Preliminar. Mexico City: ECLA.
de Murias, Ramon
1989 The Economic Regulation of International Air Transport. Jefferson NC: McFarland and Company.
Diario Oficial
1947 Ley Que Crea la Comisión Nacional de Turismo. Diario Oficial (November 25).
1980 Plan Nacional De Turismo. Diario Oficial (February 4).
1973 Ley Para Promover la Inversión Mexicana y Regular la Inversión Extranjera. Diario Oficial (March 9).
Doganis, Rigas
1993 The Bilateral Regime for Air Transport: Current Position and Future Prospects. *In* International Air Transport: The Challenges Ahead, pp. 45–73. Paris: OECD.
Dunning, John H.
1970 Trade, Location of Economic Activity and the Multinational Enterprise: A Search for an Eclectic Approach. *In* The International Allocation of Economic Activity, B. Ohlin, P. O. Hesselborn, and P. J. Wiskman, eds, pp. 395–418. London: Macmillan.
Dunning, John H., and Matthew McQueen
1982 Multinational Corporations in the International Hotel Industry. Annals of Tourism Research 9:69–90.
Edgell, David. L.
1990 International Tourism Policy. New York: Van Nostrand Reinhold.
EIU
1986 Barriers to International Travel. Travel and Tourism Analyst (3):3–13.
1987 Hotel Chains in the USA: Review of an Industry in Transition. Travel and Tourism Analyst (3):43–53.
1988 Key Problems and Prospects in the International Hotel Industry. Travel and Tourism Analyst (1):27–49.
1993 Mexico. International Tourism Reports (1):5–22.
1999 Mexico. International Tourism Reports (3):46–73.
Elite Turística
1992 Elite Turistica (July–August):18.
Ellinson, Christopher, and Gary Gereffi
1990 Explaining Strategies and Patterns of Industrial Development. *In* Manufacturing Miracles: Paths of Industrialization in Latin America and East Asia, G. Gereffi and D. Wyman, eds, pp. 368–403. Princeton NJ: Princeton University Press.
Encarnation, Dennis
1989 Dislodging Multinationals: India's Strategy in Comparative Perspective. Ithaca NY: Cornell University Press.
Enloe, Cynthia
1989 Bananas, Beaches and Bases: Making Feminist Sense of International Politics. Berkeley: University of California Press.
Enríquez Savignac, Antonio
1988 Speech to the American Chamber of Commerce of Mexico. Business Mexico 5(1):66–69.

Erfani, Julie
  1995   The Paradox of the Mexican State: Rereading Sovereignty from Independence
       to Nafta. Boulder CO: Westview.
Escobar, Arturo
  1995   Encountering Development. Princeton NJ: Princeton University Press.
Evans, Nancy
  1979   The Dynamics of Tourism Development in Puerto Vallarta. *In* Tourism: Passport
       to Development?, E. de Kadt., ed., pp. 305–320. New York: Oxford University Press.
Evans, Peter B.
  1979   Dependent Development: The Alliance of Multinational, State, and Local
       Capital in Brazil. Princeton NJ: Princeton University Press.
  1992   The State as Problem and Solution: Predation, Embedded Autonomy, and
       Structural Change. *In* The Politics of Economic Adjustment: International
       Constraints, Distributive Conflicts and the State, S. Haggard and R. R. Kaufman,
       eds, pp. 139–181. Princeton NJ: Princeton University Press.
  1995   Embedded Autonomy: States and Industrial Transformation. Princeton NJ:
       Princeton University Press.
Evans, P. B., D. Rueschemeyer, and T. Skocpol
  1985   On the Road Toward a More Adequate Understanding of the State. *In* Bringing
       the State Back, P. B. Evans, D. Rueschemeyer and T. Skocpol, eds, pp. 347–366.
       Cambridge: Cambridge University Press.
Evans, Peter B., Dietrich Rueschemeyer, and Theda Skocpol, eds.
  1985   Bringing the State Back In. Cambridge: Cambridge University Press.
Evans, Peter B., and Gary Gereffi
  1982   Foreign Investment and Dependent Development: Comparing Brazil and Mexico.
       In Brazil and Mexico: Patterns in Late Development, S. Hewlett and R. Weinert, eds,
       pp. 111–168. Philadelphia PA: Institute for the Study of Human Issues.
Expansión
  1992   Las Empresas más Importantes de México 24(527).
Fajnzylber, Fernando, and Trinidad Martínez
  1976   Las Empresas Transnacionales: Expansión a Nivel Mundial y Proyección en la
       Industria Mexicana. Mexico City: Fondo de Cultura Económica.
Feldman, Joan
  1987   International Airlines: Crossing the Border. Travel and Tourism Analyst
       (November):3–14.
El Financiero
  1992a   El Financiero (August 26):14.
  1992b   El Financiero (August 11):13.
  1992c   El Financiero (August 18):6a.
  1992d   El Financiero (July 28):11.
  1992e   El Financiero (July 2):12.
Findlay, Chris
  1990   Air Transport. *In* The Uruguay Round: Services in the World Economy, Patrick
       Messerlin and Karl P. Sauvant, eds, pp. 77–83. Washington DC: World Bank.
Findlay, Ronald
  1991   The New Political Economy: Its Explanatory Power for LDCs. *In* Politics and
       Policy Making in Developing Countries: Perspectives on the New Political
       Economy, G. M. Meier, ed., pp. 77–83. San Francisco: ICS Press.

Fitzgerald, E. V. K.
1985   The Financial Constraint on Relative Autonomy: The State and Capital Accumulation in Mexico, 1940–1982. *In* The State and Capital Accumulation in Latin America: Vol. 1: Brazil, Chile, Mexico, C. Anglade and C. Fortín, eds, pp. 210–235. Pittsburgh: University of Pittsburgh Press.

FONATUR
1990   Evaluación Económica y Financiera Ex-Post: Cancún, Q.R. Mexico City: FONATUR.
1985a   Estadísticas de Financiamiento a la Actividad Turística. Mexico City: FONATUR.
1985b   Participación de FONATUR como Agente Principal en la Creación y Desarrollo de los Centros Turísticos Integrales, 1985. Mexico City: FONATUR.
1992   Bahías de Huatulco: Magaproyectos Fonatur. Mexico City: Mexican Government Tourism Office.
nd-a   No Date-a Megaprojects Internal document Mexico City: Mexican Government Tourism Office.
n.d-b   Resumen Estadístico de Financiamiento a la Actividad Turística, 1974–1991. Internal document. Mexico City: Mexican Government Tourism Office.
n.d-c   Reglas de Operación de Crédito. Mexico City: FONATUR.
n.d-d   Transportación Aérea Requerida por los Desarrollos Turísticos de FONATUR, 1981–1984. Internal document. Mexico City: FONATUR.

Forbes
1989   Mexico Contrarian 144(5):88–89.
1994a   Evolution of a Dinasaur 154(13):264–267.
1994b   The 500 Largest Foreign Companies 154(3):228–261.
1995   Marriott, Meet Marriott 155(6): 48–50.

Frank, Andre Gunder
1967   Capitalism and Underdevelopment in Latin America. New York: Monthly Review Press

García de Fuentes, Ana
1979   Cancún: Turismo y Subdesarrollo Regional. Mexico City: Instituto de Geografía, UNAM.

Geddes, Barbara
1993   Politician's Dilemma: Building State Capacity in Latin America. Berkeley: University of California Press.

Gereffi, Gary
1983   The Pharmaceutical Industry and Dependency in the Third World. Princeton NJ: Princeton University Press.
1990   Paths of Industrialization: An Overview. *In* Manufacturing Miracles: Paths of Industrialization in Latin America and East Asia, G. Gereffi and D. L. Wyman, eds, pp. 3–31. Princeton NJ: Princeton University Press.
1994   "Contending Perspectives on Regional Integration: Development Strategies and Commodity Chains in Latin American and East Asia. Paper presented at the 18th International Congress of the Latin American Studies Association, Atlanta GA.

Gereffi, Gary, and Peter B. Evans
1981   Transnational Corporations, Dependent Development, and State Policy in the Semi-Periphery: A Comparison of Brazil and Mexico. Latin American Research Review 16(3):31–64.

Gereffi, Gary, and Miguel Korzeniewicz, eds.
1994   Commodity Chains and Global Capitalism. Westport CN: Greenwood.
Gereffi, Gary, and Donald Wyman
1989   Determinants of Development Strategies in Latin America and East Asia. *In* Pacific Dynamics: The International Politics of Industrial Change, S. Haggard and C. Moon, eds, pp. 23–52. Boulder CO: Westview.
Giarini, Orio, ed.
1987   The Emerging Service Economy. Oxford: Pergamon.
Gibbs, Murray
1987   Trade in Services: A Challenge for Development. *In* The Emerging Service Economy, O. Giarini, ed., pp. 83–104. Oxford: Pergamon.
Glade, William P., Jr.
1963   Revolution and Economic Development: A Mexican Reprise. *In* The Political Economy of Mexico, W. P. Glade and C. W. Anderson, eds, pp. 1–101. Madison WI: University of Wisconsin Press.
Golich, Vicki L.
1990   Liberalizing International Air Transport Services. *In* Privatization and Deregulation in Global Perspective, D. J. Gayle and J. N. Goodrich, eds, pp. 156–76. New York: Quorum Books.
1992   From Competition to Collaboration: The Challenge of Commercial-class Aircraft Manufacturing. International Organization 46:899–934.
Gormsen, Erdman
1989   El Turismo Internacional como Nuevo "Frente Pionero" en los Paises Tropicales. *In* Teoría y Praxis del Espacio Turístico, D. Hiernaux, ed., pp. 75–91. Mexico City: UAM-Xochimilco.
Graham, Douglas H.
1982   Mexico and Brazilian Economic Development: Legacies, Patterns and Performance. *In* Brazil and Mexico: Patterns of Late Development, S. A. Hewlett and R. S. Weinert, eds, pp. 13–55. Philadelphia: Institute for the Study of Human Issues.
Gray, H. Peter
1970   International Tourism — International Trade. Lexington MA: Heath Lexington Books.
Grieco, Joseph M.
1984   Between Dependency and Autonomy: India's Experience with the International Computer Industry. Berkeley: University of California Press.
Grindle, Merilee S.
1977   Bureaucrats, Politicians, and Peasants in Mexico. Berkeley: University of California Press.
1986   State and Countryside: Development Policy and Agrarian Politics in Latin America. Baltimore: The Johns Hopkins University Press.
1991   The New Political Economy: Positive Economics and Negative Politics. *In* Politics and Policy Making in Developing Countries, G. M. Meier, ed., pp. 41–68. San Francisco: ICS Press.
Grindle, Merilee S., and John W. Thomas
1991   Public Choices and Policy Change: The Political Economy of Reform in Developing Countries. Baltimore: The Johns Hopkins University Press.

Haber, Stephen H.
1989   Industry and Underdevelopment: The Industrialization of Mexico, 1890–1940. Stanford CA: Stanford University Press.
Haggard, Stephan
1989a   Introduction: The International Politics of Industrial Change. *In* Pacific Dynamics: The International Politics of Industrial Change, S. Haggard and C. Moon, eds, pp. 1–21. Boulder CO: Westview.
1989b   The Political Economy of Foreign Direct Investment in Latin America. Latin American Research Review 24:184–208.
1990   Pathways from The Periphery: The Politics of Growth in the Newly Industrializing Countries. Ithaca NY: Cornell University Press.
Haggard, Stephan, and Chung-in Moon
1990   Institutions and Economic Policy: Theory and a Korean Case Study. World Politics 42(1):210–237.
Hamilton, Nora
1982   The Limits of State Autonomy: Post-Revolutionary Mexico. Princeton NJ: Princeton University Press.
1986   State-Class Alliances and Conflicts: Issues and Actors in the Mexican Economic Crisis. *In* Mexico: State, Economy and Social Conflict, N. Hamilton and T. F. Harding, eds, pp. 148–174. Berverly Hills CA: Sage.
Hansen, Roger D.
1971   The Politics of Mexican Development. Baltimore: The Johns Hopkins University Press.
Hanson, Gordon H.
1994   Antitrust in Post-privatization Latin America: An Analysis of the Mexican Airlines Industry. The Quarterly Review of Economics and Finance 54(3):199–216.
Harrison, David
1992   International Tourism and the Less Developed Countries: The Background. *In* Tourism and The Less Developed Countries, D. Harrison, ed., pp. 1–34. London: Belhaven.
Hellman, Judith Adler
1983   Mexico in Crisis (2nd ed.). New York: Holmes and Meier.
Hiernaux Nicolas, D.
1989   Mitos y Realidades del Milagro Turístico. *In* Teoría y Praxis del Espacio Turístico, D. Hiernaux, ed., pp. 109–20. Mexico City: UAM-Xochimilco.
1999   Cancún Bliss. *In* The Tourist City, D. R. Judd and S. S. Fainstein, eds, pp. 124–142. New Haven CT: Yale University Press.
Hiernaux Nicolas, Daniel, and Manuel Rodríguez Woog
1991   Tourism and Absorption of the Labor Force in Mexico. *In* Regional and Sectoral Development in Mexico as Alternatives to Migration, S. Díaz Briquets and S. Weintraub, eds, pp. 311–29. Boulder CO: Westview.
Hilton, Conrad
1957   Be My Guest. Englewood Cliffs NJ: Prentice Hall.
Hoekman, Bernard M.
1990   Services-Related Production, Employment, Trade, and Factor Movements. *In* The Uruguay Round, P. A. Messerlin and K. P. Sauvant, eds, pp. 27–46. Washington DC and New York: The World Bank and UNCTC.

Holloway, J. Christopher
1989   The Business of Tourism (3rd ed.). London: Pittman.
Hotels
1994   Mexican Group Bids on Westin 28(4):3.
1998   Chasing Potential in Latin America 32(10):42–56.
1999   Hotels' Corporate 300 Ranking 33(7):50–68.
Hymer, Stephen
1976.   The International Operations of National Firms: A Study of Direct Foreign Investment. Cambridge MA: MIT Press.
IATA
1993   World Air Transport Statistics: IATA 1994. Montreal: International Air Transport Association.
1999   World Air Transport Statistics: IATA 1999. Montreal: IATA.
ICA
nd   Informe: División Turismo. Internal Document. Mexico City: Ingenieros Civiles Asociados.
IDB
Various years   Annual Report. Washington DC: Inter-American Development Bank.
IMF
Various years   International Financial Statistics. Washington DC: International Monetary Fund.
IMIT
1980   Historia del Turismo Moderno en México. Report to Consejo Nacional de Turismo. Mexico City: Instituto Mexicano de Investigaciones Turísticas.
INFRATUR/BANXICO
nd   Tourist Development Project of Cancún, Q.R. Mexico City: Banxico.
INFRATUR
nd   Basic Facts of the Tourist Project of Cancún. Mexico City: INFRATUR.
Islas Guzmán, Antonio
1989   El Caso Ixtapa-Zihuatanejo. In Teoría y Praxis del Espacio Turístico, D. Hiernaux, ed., pp. 93–108. Mexico City: UAM-Xochimilco.
Izquierdo, Rafael
1964   Protectionism in Mexico. In Public Policy and Private Enterprise in Mexico, R. Vernon, ed., pp. 243–289. Cambridge MA: Harvard University Press.
Jenkins, Barbara
1992   The Paradox of Continental Production: National Investment Policies in North America. Ithaca NY: Cornell University Press.
Jiménez Martínez, Alfonso de Jesús
1990   Turismo: Estructura y Desarrollo. Mexico City: McGraw-Hill.
Johnson, Peter
1993   Air Transport. In European Industries: Structure, Conduct, Performance, P. Johnson, ed., pp. 204–229. Aldershot: Edward Elgar.
Jönsson, Christer
1981   Sphere of Flying: The Politics of International Aviation. International Organization 35:273–302.
La Jornada
1992a   La Jornada (August 8):27.
1992b   La Jornada (August 3):36.

1992c   La Jornada (May 31):25.

Jud, G. Donald
1974   Tourism and Economic Growth in Mexico Since 1950. Inter-American Economic Affairs 28:19–43.

Kaufman, Robert R.
1990   How Societies Change Development Models or Keep Them: Reflections on the Latin American Experience in the 1930s and the Postwar World. *In* Manufacturing Miracles: Paths of Industrialization in Latin America and East Asia, G. Gereffi and D. L. Wyman, eds, pp. 110–138. Princeton NJ: Princeton University Press.

Kenen, Peter B.
1994   The International Economy, 3rd ed. Cambridge: Cambridge University Press.

Knibb, David
1996   Mexican Standoff: Aeromexico and Mexicana have Common Owner After Restructuring. Reed Business Publishing: Airline Business 12(12):9.

Knowles, Michael
1990   Packaging the Tourism Product. *In* Horwath Book of Tourism, M. Quest, ed., pp. 136–144. London: Macmillan.

Krasner, Stephen D.
1985.   Structural Conflict: The Third World Against Global Liberalism. Berkeley: University of California Press.

Krueger, Anne O.
1974   The Political Economy of the Rent-Seeking Society. American Economic Review 291–303.
1990   Government Failures in Development. Journal of Economic Perspectives 4(2):9–23.

La Botz, Dan
1992   Mask of Democracy: Labor Suppression in Mexico Today. Boston: South End Press.

Lal, Deepak
1985   The Poverty of Developmental Economics. Cambridge: Harvard University Press.

Lanfant, Marie-Françoise
1980   Introduction: Tourism in the Process of Internationalization. International Social Science Journal 32(1):14–43.

Lanvin, Bruno, ed.
1993   Trading in a New World Order: The Impact of Telecommunications and Data Services on International Trade in Services. Boulder CO: Westview.

Latin American Weekly Report
1988a   WR-88-47. Latin American Weekly Report (December 1):5.
1988b   WR-88-27. Latin American Weekly Report (July 14):12.

LatinFinance
1990   One More Time. LatinFinance 3(16):30–32.

Lattin, Thomas W.
1990   Hotel Technology: Key to Survival. *In* Horwath Book of Tourism, M. Quest, ed., pp. 219–223. London: Macmillan.

Lea, John.
1988   Tourism and Development in the Third World. London: Routledge.

Leary, W. M.
1992   The Airline Industry. New York: Facts on File.

Levin, Baron F.
 1999   Clear For Takeoff: Mexico's Aviation Industry Gaining Altitude Following Period of Turbulence. Business Mexico 9(4):32–34.
Lindblom, Charles
 1977   Politics and Markets. New York: Basic Books.
Long, Veronica
 1991   Government — Industry — Community Interaction in Tourism Development in Mexico. *In* The Tourism Industry: An International Analysis, M. T. Sinclair and M. J. Stabler, eds, pp. 205–222. Wallingford: CAB International.
Looney, Robert E.
 1978   Mexico's Economy: A Policy Analysis with Forecasts to 1990. Boulder CO: Westview.
Luna Ledesma, Matilde
 1992   Los Empresarios y el Cambio Político: México, 1970–1987. Mexico City: Instituto de Investigaciones Sociales, UNAM/Era.
Lundberg, Donald E., Mink H. Stavenga, and M. Krishnamoorthy
 1995   Tourism Economics. New York: J. Wiley.
Lustig, Nora
 1992   Mexico: The Remaking of an Economy. Washington DC: The Brookings Institution.
MacDonald Escobedo, Eugenio
 1981   Turismo: Una Recapitulación. Mexico City: Editorial Bodini.
Madeley, John
 1992   Trade and the Poor: The Impact of International Trade on Developing Countries. New York: St. Martin's Press.
Madsen Camacho, Michelle E.
 1996   Dissenting Workers and Social Control: A Case Study of the Hotel Industry in Huatulco, Oaxaca. Human Organization 55(1):32–40.
Mattelart, Armand
 1974   La Cultura Como Empresa Multinacional. Mexico City: Era.
Matthews, Harry
 1978   International Tourism: A Political and Social Analysis. Cambridge: Schenkman Publishing.
Maxfield, Sylvia
 1990   Governing Capital: International Finance and Mexican Politics. Ithaca NY: Cornell University Press.
 1991   Bankers' Aliances and Economic Policy Patterns: Evidence from Mexico and Brazil. Comparative Political Studies 23:419–458.
Meneses Gómez, Esteben
 1987   El Contado Público en la Industria Hotelera y su Participación en la Economía Nacional. Unpublished Thesis, Instituto Politécnico Nacional.
Meyer, Lorenzo
 1977   Mexico and the United States in the Oil Controversy, 1917–1942. Austin: University of Texas Press.
Molinero Molinero, Rosario Asela.
 1983   Mitos y Realidades de Turismo en Mexico (1976–1981). Masters thesis in tourism, El Colegio de Mexico.

Moran, Theodore
1974 Multinational Corporations and the Politics of Dependence: Copper in Chile. Princeton NJ: Princeton University Press.
Moreno Toscano, Octavio
1970 La Estructura Internacional del Negocio Turístico. Comercio Exterior 20:3.
1971 El Turismo com Factor Político en las Relaciones Internacionales. Foro Internacional 12(1):66–94.
Morgan Stanley
nd Overview of Selected Lodging Companies and Their Partners. Internal Document, available from the author.
Morrison, Steven A., and Clifford Winston
1995 The Evolution of the Airline Industry. Washington DC: The Brookings Institution.
Mosk, Sanford A.
1954 Industrial Revolution in Mexico. Berkeley: University of California Press.
Munguía Huato, Romàn
1989 Los Empresarios de la Obra Pública en México. El Capital Monopolista en la Construcción. In Empresarios de México: Aspectos Históricos, Económicos e Ideológicos, E. Jacobo, M. Luna y R. Tirado, eds, pp. 193–222. Guadalajara: Universidad de Guadalajara.
El Nacional
1992 Corporación Falcón Interesada por Capital Social de CMA. El Nacional (August 22):26.
NAFINSA
Various years La Economía Mexicana en Cifras. Mexico City: Nacional Financiera.
1966 Statistics on the Mexican Economy. Mexico City: NAFINSA.
1971 Fondo de Garantía y Fomento del Turismo. Internal document. Mexico City: NAFISA.
Newfarmer, Richard S.
1985 International Industrial Organization and Development: A Survey. In Profits, Progress and Poverty: Case Studies of International Industries in Latin America, R. S. Newfarmer, ed., pp. 13–61. Notre Dame IN: University of Notre Dame Press.
New York Times
1972 Why the Computer Chose Cancun. New York Times (March 5, Section 10):1.
1976 The Story behind the Mexico Boycott. New York Times (June 27, Section 10):1.
1988 Mexico Air Travel in Turmoil. New York Times (April 25):D-8.
1991 Mexico Courts the Bus Tourists. New York Times (November 30):35.
1992a Bancomer to Manage Hotels in Cancun and Puerto Vallanta (advertisement). New York Times (December 2):D-14.
1992b Mexico's Little Airline that Could. New York Times (November 13):D-1.
1995a The Gold in Latin Skies. New York Times (August 26):34.
1995b Ramada is Revamping Hotels to Raise Profit Margins. New York Times (April 7):D-4.
1995c Stretching Your Money in Mexico. New York Times (February 19):14.
1995d A Default Worsens Mexico's Ills. New York Times (February 16):D-1.
1995e Tons of Cocaine Leaving Mexico in Old Jets. New York Times (January 10):A-1.

Nordlinger, Eric
1981 On the Autonomy of the Democratic State. Cambridge: Harvard University Press.
O'Connor, William E.
1989 An Introduction to Airline Economics (4th ed). Westport CO: Praeger.
O'Hearn, Denis.
1990 The Road from Import-Substitution to Export-Led Industrialization in Ireland: Who Mixed the Asphalt, Who Drove the Machinery, and Who Kept Making Them Change Directions? Politics and Society 18(1):1–38.
OAS
1979 Hemispheric Policy on Tourism Development and Strategy for Implementation. Washington DC: Organization of American States.
OECD
1989 Trade in Services and Developing Countries. Paris: Organization for Economic Cooperation and Development.
1993 International Air Transport: The Challenges Ahead. Paris: Organization for Economic Cooperation and Development.
Orr, Bill
1992 The Global Economy in the 90s: A User's Guide. New York: New York University Press.
Ortiz de la Peña Rodríguez, Oscar E.
1981 La Política de Promoción Fiscal al Turismo.Professional thesis in tourism studies, Intituto Tecnológico Autónomo de México.
Page, John M.
1994 The East Asian Miracle: An Introduction. World Development 22(4):615–625.
Pearce, Douglas
1989 Tourist Development (2nd ed.). Essex: Longman.
Peres Nuñez, Wilson
1990 Foreign Investment and Industrial Development in Mexico. Paris: OECD.
Perrin, Dennis
1986 La Hotelería. Mexico City: Fondo de Cultura Económica.
Petzinger, Thomas
1995 Hard Landing. New York: Times Books.
Pi-Sunyer, Oriol, and R. Brooke Thomas
1997 Tourism, Environmentalism, and Cultural Survival in Quintana Roo. *In* Life and Death Matters: Human Rights and The Environment at the End of the Millenium, B. Rose Johnston, ed., pp. 187–212. Walnut Creek CA: AltaMira Press.
PKF
1991 Tendencias en la Hotelería Mexicana. Mexico City: PKF.
Poon, Auliana
1990 Felxible Specialization and Small Size: The Case of Caribbean Tourism. World Development 18(1):109–123.
Poulantzas, Nicos
1978. State, Power, Socialism. London: New Left Books.
Proceso
1991 En Los Cabos, Hotels, Residencias, Siele Campos de Golf, Cinco Kilómetros de Maya. Proceso 765 (July 1):22–25.

1994 Hank y sus 42 años en la vida públita: la politica como negocio, la adulación como instrumento. Proceso 941 (November 14):18–23.

Ram, Rati
1987 Exports and Economic Growth in Developing Countries: Evidence from Time-Series and Cross-Section Data. Economic Development and Cultural Change 36(1):51–72.

Ramírez Blanco, Hector Manuel
1983 Teoría General de Turismo. Mexico City: Diana.

Ramírez Saiz, Juan Manuel
1989 Turismo y Medio Ambiente: El Caso de Acapulco. *In* Teoría y Praxis del Espacio Turístico, D. Hiernaux Nicolás, ed., pp. 35–70. Mexico City:UAM-Xochimilco.

Randall, Clarence B.
1958 Report to the President of the United States: International Travel. Washington DC: US Government Printing Office.

RCI de México
1992 Estadísticas Básicas del Tiempo Compartido en México. Internal document. Mexico City: RCI de México.

Reid, Gavin
1987 Theories of Industrial Organization. Oxford: Blackwell.

República de México
nd Resumen de la Propuesta de Préstamo a Nacional Financiera, S.A., Proyecto de Desarrollo Turístico Cancún. Internal Document. Mexico City: FONATUR.

Reynolds, Clark W.
1970 The Mexican Economy: Twentieth Century Structure and Growth. New Haven CN: Yale University Press.
1978 Why Mexico's "Stabilizing Development" Was Actually Destabilizing (With Some Implications for the Future). World Development 6:1005–1018.

Reynosa y Valle Augustín, and Jacomina P. De Regt
1979 Growing Pains: Planned Tourism Development in Ixtapa-Zihuatanejo. *In* Tourism: Passport to Development?, E. de Kadt, ed., pp. 111–134. New York: Oxford University Press.

Richter, Christine
1987 Tourism Services. *In* The Emerging Services Economy, O. Giarini, ed., pp. 213–244. Oxford: Pergamon.

Richter, Linda K.
1983 Tourism Politics and Political Science: A Case of not so Benign Neglect. Annals of Tourism Research 10:313–335.
1989 The Politics of Tourism in Asia. Honolulu: University of Hawaii Press.

Riddle, Dorothy I.
1986 Service-led Growth: The Role of the Service Sector in World Development. New York: Praeger.

Rivera Corona, Joel, and Francisco E. Ron Delgado
1991 Agencias de Viajes y Operadores de Excursiones en México. *In* México: Una Economía de Servicios, UNCTAD/SECOFI, eds, pp. 199–216. New York: UN.

Rubio, Luis, Edna Jaime, and Alberto Díaz
1990 Mexico. *In* The Uruguay Round: Services in the World Economy, P. A. Messerlin and K. Sauvant, eds, pp. 166–174. Washington DC: World Bank.

Sachs, Wolfgang, ed.
1992   The Development Dictionary: A Guide to Knowledge as Power. London: Zed.
Sánchez, Manuel, Rossana Corona, Otoniel Ochoa, Luis Fernando Herrera, Arturo Olvera and Ernesto Sepúlveda
1993   The Privatization Process in Mexico: Five Case Studies. *In* Privatization in Latin America, M. Sánchez and R. Corona, eds, pp. 101–119. Washington DC: Inter-American Development Bank.
Saragoza, Alex
1988   The Monterrey Elite and the Mexican State, 1880–1940. Austin: University of Texas Press.
Schédler, Andreas
1988   El Capital Extranjero en México: El Caso de la Hotelería. Investigación Económica 184:137–175.
Schlüter, Regina
1994   Tourism Development: A Latin American Perspective. *In* Global Tourism: The Next Decade, W. F. Theobald, ed., pp. 246–260. Oxford: Butterworth Heinemann.
SCT
1990   La Aviación Mexicana en Cifras, 1980–1989. Mexico City: SCT.
1991   El Trasporte Aéreo de Pasajeros y Carga, 1989–1990. Mexico City: SCT.
SECTUR
Various years   Informe de Labores. Mexico City: SECTUR.
1983   Memoria de Labores, 1976–1982. Mexico City: SECTUR.
1985   Estadísticas Oportunas Sobre Turismo. Mexico City: SECTUR.
1986   Carpeta Estadística de Turismo. Mexico City: SECTUR.
1989   Programa Nacional de Modernización del Turismo, 1989–1994. Mexico City: Poder Ejecutivo Federal.
1991a   Mexico's Tourism Sector: The Year in Review, 1990. Internal document. Mexico City: SECTUR.
1991b   Programa Nacional de Modernización del Turismo, 1991–1994. Mexico City: Poder Ejecutivo Federal.
1991c   The Trust Mechanism in Tourism Investment in Mexico. Mexico City: SECTUR.
1992   Estadísticas Básicas de la Actividad Turística. Mexico City: Sectur/Bancomer.
nd   Evolución de la Actividad Turística 1970–1991. Mexico City: SECTUR.
SECTUR/FONATUR
nd   Megaprojects. Mexico City: SECTUR/FONATUR
Shacochis, Bob
1989   In Deepest Gringolandia. Harpers 279(1670):42–50.
Shafer, D. Michael
1994   Winners and Losers: How Sectors Shape the Developmental Prospects of States. Ithaca NY: Cornell University Press.
Shapiro, Helen
1994   Engines of Growth: The State and Transnational Auto Companies in Brazil. New York: Cambridge University Press.
Shelton, David
1964   The Banking System: Money and the Goal of Growth. *In* Public Policy and Private Enterprise in Mexico, R. Vernon, ed. Cambridge: Harvard University Press.

Shrestha, Nanda
1995 Becoming a Development Category. *In* Power of Development, J. Crush, ed., pp. 226–277. London: Routledge.
Sikkink, Kathryn
1991 Ideas and Institutions: Developmentalism in Brazil and Argentina. Ithaca NY: Cornell University Press.
Sinclair, M. Thea, Parvin Alizadeh, Elizabeth Atieno, and Adero Aonunga
1992 The Structure of International Tourism and Tourism Development in Kenya. *In* Tourism and the Less Developed Countries, D. Harrison, ed., pp. 47–63. London: Belhaven.
Skocpol, Theda
1985 Bringing the State Back In: Strategies of Analysis in Current Research. In Bringing the State Back In, P. B. Evans, D. Rueschemeyer and T. Skocpol, eds, pp. 3–43. Cambridge: Cambridge University Press.
Smith, Peter H.
1991 Mexico Since 1946: Dynamics of an Authoritarian Regime. *In* Mexico Since Independence, L. Bethell, ed., pp. 321–396. New York: Cambridge University Press.
Snape, Richard H.
1990 Principles in Trade in Services. *In* The Uruguay Round: Services in the World Economy, P. A. Messerlin and K. P. Sauvant, eds, pp. 5–11. Washington DC: World Bank.
Sochor, Eugene
1991 The Politics of International Aviation. Iowa City: University of Iowa Press.
Solís, Leopoldo
1971 La Realidad Económica Mexicana: Retrovisión y Perspectivas. Mexico City: Siglo Veintiuno.
1981 Economic Policy Reform in Mexico: A Case Study for Developing Countries. New York: Pergamon.
SPP
1985 Informe Sobre las Razones y Criterios que Fundamentan las Medidas de Reestructuración de la Administración Pública Federal Paraestatal. Anexos. Mexico City: SPP.
Stallings, Barbara
1978 Class Conflict and Economic Development in Chile, 1958–1973. Stanford CA: Stanford University Press.
1995 Introduction: Global Change, Regional Response. *In* Global Change, Regional Response: The New International Context of Development, B. Stallings, ed., pp. 1–30. New York: Cambridge University Press.
Sussman, Silvia, and Chris Cooper
1993 Latin America and Africa. *In* The International Hospitality Industry: Organizational and Operational Issues, P. Jones and A. Pizam, eds, pp. 68–85. New York: Wiley.
Tancer, Robert S.
1972 Tourist Promotion in Mexico. Law and the Social Order (4):559–579.
1975 Tourism in the Americas: Some Governmental Initiatives, Paper presented at the second Arizona Latin America Conference. Center for Latin American Studies, Arizona State University.

Taneja, Nawal K.
  1988   The International Airline Industry: Trends, Issues and Challenges. Lexington MA: Lexington Books.
Taylor, Lance
  1986   Trade and Growth. The Review of Black Political Economy (Spring):17–36.
Teichman, Judith A.
  1988   Policymaking in Mexico: From Boom to Crisis. Boston: Allen Unwin.
Thomas, Barry
  1993   Tourism. *In* European Industries: Structure, Conduct, Performance, P. Johnson, ed., pp. 230–252. Aldershot: Edward Elgar.
Torruco Marqués, Miguel
  1988   Historia Institucional del Turismo en México, 1926–1988. Unpublished paper.
Travel and Tourism Intelligence
  1999   Mexico. TTI Country Reports. Travel and Tourism Intelligence (3):50–73.
Truett, Lila J., and Dale B. Truett
  1982   Public Policy and the Growth of the Mexican Tourism Industry, 1970–1970. Journal of Travel Research 20(4):11–19.
Turner, Louis
  1976   The International Division of Leisure: Tourism and the Third World. World Development 4(3):253–260.
Turner, Louis, and John Ash
  1975   The Golden Hordes: International Tourism and the Pleasure Periphery. London: Constable.
United Nations
  1963   Recommendations on International Tourism. Rome: UN Conference on Tourism. New York: United Nations.
  1973   Elements of Tourism Policy in Developing Countries. New York: United Nations.
UNCTC
  1982   Transnational Corporations in International Tourism. New York: United Nations Centre on Transnational Corporations.
  1990   Transnational Corporations, Services, and the Uruguay Round. New York: United Nations Centre on Transnational Corporations.
United Nations Transnational Corporations and Management Division, Department of Economic and Social Development
  1993   The Transnationalization of Service Industries. New York: United Nations.
United States Treaties and Other International Agreements
  nd   TIAS 4675, pp. 60–77. Washington DC: US Government Printing Office.
Unomásuno
  1992   Unomásuno (July 19, supplement):5.
Urry, John
  1990   The Tourist Gaze: Leisure and Travel in Contemporary Societies. London: Sage.
van den Berghe, Pierre L.
  1994   The Quest for the Other: Ethnic Tourism in San Cristóbal, Mexico. Seattle: University of Washington Press.
Var, Turgut, John Ap, and Carlton Van Doren
  1994   Tourism and World Peace. *In* Global Tourism: The Next Decade, W. F. Theobald, ed., pp. 27–39. Oxford: Butterworth-Heinemann.

Vernon, Raymond
1971   Sovereignty at Bay: The Multinational Spread of U.S. Enterprises. New York: Basic Books.
Vernon, Raymond, ed.
1964   Public Policy and Private Enterprise in Mexico. Cambridge: Harvard University Press.
Vietor, Richard H. K.
1994   Contrived Competition: Regulation and Deregulation in America. Cambridge MA: The Belnap Press of Harvard University.
Villareal, René.
1990   The Latin American Strategy of Import Substitution: Failure or Paradigm for the Region? *In* Manufacturing Miracles: Paths of Industrialization in Latin America and East Asia, G. Gereffi and D. L. Wyman, eds, pp. 292–320. Princeton: Princeton University Press.
Wade, Robert
1990a   Governing the Market: Economic Theory and the Role of Government in East Asian Dynamism. Princeton NJ: Princeton University Press.
1990b   Industrial Policy in East Asia: Does it Lead or Follow the Market? *In* Manufacturing Miracles: Paths of Industrialization in Latin America and East Asia, G. Gereffi and D. L. Wyman, eds, pp. 231–266. Princeton NJ: Princeton University Press.
Waterbury, John
1992   The Heart of the Matter? Public Enterprise and the Adjustment Process. *In* The Politics of Economic Adjustment: International Constraints, Distributive Conflicts, and the State, S. Haggard and R. R. Kaufman, eds, pp. 182–217. Princeton NJ: Princeton University Press.
Whiting, Van R., Jr.
1992   The Political Economy of Foreign Investment in Mexico: Nationalism, Liberalism, and Constraints on Choice. Baltimore: The Johns Hopkins University Press.
Williams, Allen, and Gareth Shaw
1991   Introduction. *In* Tourism and Economic Development: Western European Experiences. London: Belhaven Press.
Williamson, John
1990   What Washington Means by Policy Reform. *In* Latin American Adjustment: How Much has Happened?, J. Williamson, ed., pp. 5–38. Washington DC: Institute for International Economics.
Williamson, John, and Stephan Haggard
1994   The Political Conditions for Economic Reform. *In* The Political Economy of Policy Reform, J. Williamson, ed., pp. 525–596. Washington DC: Institute for International Economics.
Wilson, Patricia
1992   Exports and Local Development: Mexico's New Maquiladoras. Austin: University of Texas Press.
Witt, Stephen F., Michael Z. Brook, and Peter J. Buckley
1991   The Management of International Tourism. London: Unwin Hyman.
World Bank
1972   Tourism: Sector Working Paper. Washington DC: World Bank.
1977   News Release (77/101).

1989   Annual Report. Washington DC: World Bank.
1993   The East Asian Miracle: Economic Growth and Public Policy. New York: Oxford University Press.
1995   World Development Report, 1995. New York: Oxford University Press.
World Bank and UNCTC
1990   The Uruguay Round: Services in the World Economy. Washington DC: World Bank and UNCTC.
WTO
Various years   Yearbook of Tourism Statistics, various issues. Madrid: World Tourism Organization.

Author Index

Subject Index